INVESTIGATIONS ON THE THEORY OF THE BROWNIAN MOVEMENT

BY

ALBERT EINSTEIN, Ph.D.

EDITED WITH NOTES BY

R. FÜRTH

TRANSLATED BY

A. D. COWPER

WITH 3 DIAGRAMS

DOVER PUBLICATIONS, INC.

PREFACE

ALBERT EINSTEIN was born on 14 March, 1879, in Ulm. When he was only six weeks old his parents moved to Munich, where he spent his infancy, and went to school until his fourteenth year. When fifteen he came to Switzerland, attended for another year the Gymnasium in Aarau, and took there his school leaving examination. Then he studied Mathematics and Physics in the Zurich Polytechnic, where Minkowski was one of his teachers. In 1902 he came to Berne as Engineer in the Patent Office, and in addition to his duties there, prepared himself for the examination for his Doctor's degree, which he took in the year 1905. At this time there appeared in rapid succession his first great works on the foundations of molecular physics, of which those relating to the Brownian motions are collected in this little volume; as well as the well-known papers on the special Principle of Relativity. In 1909 he accepted a call to a Professorship in the University of Zurich, and in 1911 a call to a full Professorship in the University of Prague; in 1912 he accepted a Chair in the Zurich Polytechnic. In 1914 he was invited to Berlin as successor to Van't Hoff in the Royal Prussian Academy of Science, where in addition he undertook the duties of Director of the Kaiser Wilhelm Institute of Physics. To this period, up to the year 1915, belong his researches on the general Theory of Relativity, as well as a number of fundamental studies on the Quantum Theory.

CONTENTS

THEORY OF BROWNIAN MOVEMENT

INVESTIGATIONS ON THE THEORY OF THE BROWNIAN MOVEMENT

I

ON THE MOVEMENT OF SMALL PARTICLES SUSPENDED IN A STATIONARY LIQUID DEMANDED BY THE MOLECULAR-KINETIC THEORY OF HEAT

IN this paper it will be shown that according to the molecular-kinetic theory of heat, bodies of microscopically-visible size suspended in a liquid will perform movements of such magnitude that they can be easily observed in a microscope, on account of the molecular motions of heat. It is possible that the movements to be discussed here are identical with the so-called " Brownian molecular motion " ; however, the information available to me regarding the latter is so lacking in precision, that I can form no judgment in the matter (1).

If the movement discussed here can actually be observed (together with the laws relating to it that one would expect to find), then classical thermodynamics can no longer be looked upon as applicable with precision to bodies even of dimensions distinguishable in a microscope : an exact determination of actual atomic dimensions is then possible. On the other hand, had the prediction of this movement proved to be incorrect, a weighty argument would be provided against the molecular-kinetic conception of heat.

§ 1. On the Osmotic Pressure to be Ascribed to the Suspended Particles

Let z gram-molecules of a non-electrolyte be dissolved in a volume V^* forming part of a quantity of liquid of total volume V. If the volume V^* is separated from the pure solvent by a partition permeable for the solvent but impermeable for the solute, a so-called " osmotic pressure," p, is exerted on this partition, which satisfies the equation

$$pV^* = RTz \qquad . \qquad . \qquad . \quad (2)$$

when V^*/z is sufficiently great.

On the other hand, if small suspended particles are present in the fractional volume V^* in place of the dissolved substance, which particles are also unable to pass through the partition permeable to the solvent : according to the classical theory of thermodynamics—at least when the force of gravity (which does not interest us here) is ignored—we would not expect to find any force acting on the partition ; for according to ordinary conceptions the " free energy " of the system appears to be independent of the position of the partition and of the suspended particles, but dependent only on the total mass and qualities of the suspended material, the liquid and the partition, and on the pressure and temperature. Actually, for the calculation of the free energy the energy and entropy of the boundary-surface (surface-tension forces) should also be considered ; these can be excluded if the size and condition of the surfaces of contact do not alter with the changes in position of the partition and of the suspended particles under consideration.

But a different conception is reached from the standpoint of the molecular-kinetic theory of heat. According to this theory a dissolved molecule is differentiated from a suspended body *solely* by its dimensions, and it is not apparent why a number of suspended particles should not produce the same osmotic pressure as the same number of molecules. We must assume that the suspended particles perform an irregular movement—even if a very slow one—in the liquid, on account of the molecular movement of the liquid ; if they are prevented from leaving the volume $V*$ by the partition, they will exert a pressure on the partition just like molecules in solution. Then, if there are n suspended particles present in the volume $V*$, and therefore $n/V* = \nu$ in a unit of volume, and if neighbouring particles are sufficiently far separated, there will be a corresponding osmotic pressure p of magnitude given by

$$p = \frac{RT}{V*} \frac{n}{N} = \frac{RT}{N} \cdot \nu,$$

where N signifies the actual number of molecules contained in a gram-molecule. It will be shown in the next paragraph that the molecular-kinetic theory of heat actually leads to this wider conception of osmotic pressure.

§ 2. Osmotic Pressure from the Standpoint of the Molecular-Kinetic Theory of Heat (*)

If p_1, p_2, ... p_l are the variables of state of

(*) In this paragraph the papers of the author on the " Foundations of Thermodynamics " are assumed to be familiar to the reader (*Ann. d. Phys.*, **9**, p. 417, 1902 ; **11**, p. 170, 1903). An understanding of the conclusions reached in the present paper is not dependent on a knowledge of the former papers or of this paragraph of the present paper.

a physical system which completely define the instantaneous condition of the system (for example, the co-ordinates and velocity components of all atoms of the system), and if the complete system of the equations of change of these variables of state is given in the form

$$\frac{\partial p_\nu}{\partial t} = \phi_\nu(p_1 \ldots p_l) \ (\nu = 1, 2, \ldots l)$$

whence

$$\Sigma \frac{\partial \phi_\nu}{\partial p_\nu} = 0,$$

then the entropy of the system is given by the expression

$$S = \frac{\overline{E}}{T} + 2x \ lg \int e^{-\frac{E}{2xT}} dp_1 \ldots dp_l \qquad . \quad (3)$$

where T is the absolute temperature, $\overline{E}$ the energy of the system, E the energy as a function of p_ν. The integral is extended over all possible values of p_ν consistent with the conditions of the problem. x is connected with the constant N referred to before by the relation $2xN = R$. We obtain hence for the free energy F,

$$F = -\frac{R}{N}T \ lg \int e^{-\frac{EN}{RT}} dp_1 \ldots dp_l = -\frac{RT}{N} \ lg \ B.$$

Now let us consider a quantity of liquid enclosed in a volume V ; let there be n solute molecules (or suspended particles respectively) in the portion V^* of this volume V, which are retained in the volume V^* by a semi-permeable partition ; the integration limits of the integral B obtained in the expressions for S and F will be affected accordingly. The combined volume of the solute molecules (or suspended particles) is taken as small compared with V^*. This system will be completely defined according to the theory under discussion by the variables of condition $p_1 \ldots p_l$.

If the molecular picture were extended to deal with every single unit, the calculation of the integral B would offer such difficulties that an exact calculation of F could be scarcely contemplated. Accordingly, we need here only to know how F depends on the magnitude of the volume V^*, in which all the solute molecules, or suspended bodies (hereinafter termed briefly " particles ") are contained.

We will call x_1, y_1, z_1 the rectangular co-ordinates of the centre of gravity of the first particle, x_2, y_2, z_2 those of the second, etc., x_n, y_n, z_n those of the last particle, and allocate for the centres of gravity of the particles the indefinitely small domains of parallelopiped form dx_1, dy_1, dz_1 ; $dx_2,$

dy_2, dz_2, . . . dx_n, dy_n, dz_n, lying wholly within V^*. The value of the integral appearing in the expression for F will be sought, with the limitation that the centres of gravity of the particles lie within a domain defined in this manner. The integral can then be brought into the form

$$dB = dx_1\, dy_1 \ldots dz_n \,.\, J,$$

where J is independent of dx_1, dy_1, etc., as well as of V^*, i.e. of the position of the semi-permeable partition. But J is also independent of any special choice of the position of the domains of the centres of gravity and of the magnitude of V^*, as will be shown immediately. For if a second system were given, of indefinitely small domains of the centres of gravity of the particles, and the latter designated $dx_1'dy_1'dz_1'$; $dx_2'dy_2'dz_2'$ . . . $dx_n'dy_n'dz_n'$, which domains differ from those originally given in their position but not in their magnitude, and are similarly all contained in V^*, an analogous expression holds :—

$$dB' = dx_1'dy_1' \ldots dz_n' \,.\, J'.$$

Whence

$$dx_1 dy_1 \ldots dz_n = dx_1'dy_1' \ldots dz_n'.$$

Therefore

$$\frac{dB}{dB'} = \frac{J}{J'}$$

But from the molecular theory of Heat given in the paper quoted,(*) it is easily deduced that dB/B (4) (or dB'/B respectively) is equal to the probability that at any arbitrary moment of time the centres of gravity of the particles are included in the domains $(dx_1 \ldots dz_n)$ or $(dx_1' \ldots dz_n')$ respectively. Now, if the movements of single particles are independent of one another to a sufficient degree of approximation, if the liquid is homogeneous and exerts no force on the particles, then for equal size of domains the probability of each of the two systems will be equal, so that the following holds:

$$\frac{dB}{B} = \frac{dB'}{B}.$$

But from this and the last equation obtained it follows that

$$J = J'.$$

We have thus proved that J is independent both of V^* and of $x_1, y_1, \ldots z_n$. By integration we obtain

$$B = \int J dx_1 \ldots dz_n = J . V^* n,$$

and thence

$$F = -\frac{RT}{N}\{lg\ J + n\ lg\ V^*\}$$

(*) A. Einstein, *Ann. d. Phys.*, **11**, p. 170, 1903.

and

$$p = -\frac{\partial F}{\partial V^*} = \frac{RT}{V^*}\frac{n}{N} = \frac{RT}{N}\nu.$$

It has been shown by this analysis that the existence of an osmotic pressure can be deduced from the molecular-kinetic theory of Heat ; and that as far as osmotic pressure is concerned, solute molecules and suspended particles are, according to this theory, identical in their behaviour at great dilution.

§ 3. Theory of the Diffusion of Small Spheres in Suspension

Suppose there be suspended particles irregularly dispersed in a liquid. We will consider their state of dynamic equilibrium, on the assumption that a force K acts on the single particles, which force depends on the position, but not on the time. It will be assumed for the sake of simplicity that the force is exerted everywhere in the direction of the x axis.

Let ν be the number of suspended particles per unit volume ; then in the condition of dynamic equilibrium ν is such a function of x that the variation of the free energy vanishes for an arbitrary virtual displacement δx of the suspended substance. We have, therefore,

$$\delta F = \delta E - T\delta S = 0.$$

It will be assumed that the liquid has unit area of cross-section perpendicular to the x axis and is bounded by the planes $x = 0$ and $x = l$. We have, then,

$$\delta E = -\int_0^l K\nu\delta x\,dx$$

and

$$\delta S = \int_0^l R\frac{\nu}{N}\frac{\partial\delta x}{\partial x}dx = -\frac{R}{N}\int_0^l\frac{\partial\nu}{\partial x}\delta x\,dx.$$

The required condition of equilibrium is therefore

(1) $$-K\nu + \frac{RT}{N}\frac{\partial\nu}{\partial x} = 0$$

or

$$K\nu - \frac{\partial p}{\partial x} = 0 \quad . \quad . \quad . \quad (5)$$

The last equation states that equilibrium with the force K is brought about by osmotic pressure forces.

Equation (1) can be used to find the coefficient of diffusion of the suspended substance. We can look upon the dynamic equilibrium condition considered here as a superposition of two processes proceeding in opposite directions, namely :—

1. A movement of the suspended substance under the influence of the force K acting on each single suspended particle.

2. A process of diffusion, which is to be looked upon as a result of the irregular movement of the particles produced by the thermal molecular movement.

If the suspended particles have spherical form (radius of the sphere $= P$), and if the liquid has a coefficient of viscosity k, then the force K imparts to the single particles a velocity (*)

$$\frac{K}{6\pi kP} \qquad . \qquad . \qquad . \quad (6)$$

and there will pass a unit area per unit of time

$$\frac{\nu K}{6\pi kP}$$

particles.

If, further, D signifies the coefficient of diffusion of the suspended substance, and μ the mass of a particle, as the result of diffusion there will pass across unit area in a unit of time,

$$- D\frac{\partial(\mu\nu)}{\partial x} \text{ grams}$$

or

$$- D\frac{\partial\nu}{\partial x} \text{ particles.}$$

(*) Cf. e.g. G. Kirchhoff, "Lectures on Mechanics," Lect. 26, § 4.

Since there must be dynamic equilibrium, we must have

(2)
$$\frac{\nu K}{6\pi kP} - D\frac{\partial \nu}{\partial x} = 0.$$

We can calculate the coefficient of diffusion from the two conditions (1) and (2) found for the dynamic equilibrium. We get

$$D = \frac{RT}{N}\frac{1}{6\pi kP} \quad \cdot \quad \cdot \quad \cdot \quad (7)$$

The coefficient of diffusion of the suspended substance therefore depends (except for universal constants and the absolute temperature) only on the coefficient of viscosity of the liquid and on the size of the suspended particles.

§ 4. On the Irregular Movement of Particles Suspended in a Liquid and the Relation of this to Diffusion

We will turn now to a closer consideration of the irregular movements which arise from thermal molecular movement, and give rise to the diffusion investigated in the last paragraph.

Evidently it must be assumed that each single particle executes a movement which is independent of the movement of all other particles; the movements of one and the same particle after different intervals of time must be considered as mutually independent processes, so long as we think of these intervals of time as being chosen not too small.

We will introduce a time-interval τ in our discussion, which is to be very small compared with the observed interval of time, but, nevertheless, of such a magnitude that the movements executed by a particle in two consecutive intervals of time τ are to be considered as mutually independent phenomena (8).

Suppose there are altogether n suspended particles in a liquid. In an interval of time τ the x-co-ordinates of the single particles will increase by $\varDelta$, where $\varDelta$ has a different value (positive or negative) for each particle. For the value of $\varDelta$ a certain probability-law will hold ; the number dn of the particles which experience in the time-interval τ a displacement which lies between $\varDelta$ and $\varDelta + d\varDelta$, will be expressed by an equation of the form

$$dn = n\phi(\varDelta)d\varDelta,$$
where

$$\int_{-\infty}^{+\infty} \phi(\varDelta)d\varDelta = 1$$

and ϕ only differs from zero for very small values of $\varDelta$ and fulfils the condition

$$\phi(\varDelta) = \phi(-\varDelta).$$

We will investigate now how the coefficient of diffusion depends on ϕ, confining ourselves again to the case when the number ν of the particles per unit volume is dependent only on x and t.

Putting for the number of particles per unit volume $\nu = f(x, t)$, we will calculate the distribution of the particles at a time $t + \tau$ from the distribution at the time t. From the definition of the function $\phi(\Delta)$, there is easily obtained the number of the particles which are located at the time $t + \tau$ between two planes perpendicular to the x-axis, with abscissæ x and $x + dx$. We get

$$f(x, t + \tau)dx = dx. \int_{\Delta = -\infty}^{\Delta = +\infty} f(x + \Delta)\phi(\Delta)d\Delta.$$

Now, since τ is very small, we can put

$$f(x, t + \tau) = f(x, t) + \tau\frac{\partial f}{\partial t}.$$

Further, we can expand $f(x + \Delta, t)$ in powers of Δ :—

$$f(x+\Delta, t) = f(x, t) + \Delta\frac{\partial f(x, t)}{\partial x} + \frac{\Delta^2}{2!}\frac{\partial^2 f(x, t)}{\partial x^2} \ldots ad\ inf.$$

We can bring this expansion under the integral sign, since only very small values of Δ contribute anything to the latter. We obtain

$$f + \frac{\partial f}{\partial t} \cdot \tau = f\int_{-\infty}^{+\infty}\phi(\Delta)d\Delta + \frac{\partial x}{\partial f}\int_{-\infty}^{+\infty}\Delta\phi(\Delta)d\Delta$$
$$+ \frac{\partial^2 f}{\partial x^2}\int_{-\infty}^{+\infty}\frac{\Delta^2}{2}\phi(\Delta)d\Delta \ldots$$

On the right-hand side the second, fourth, etc., terms vanish since $\phi(x) = \phi(-x)$; whilst of the first, third, fifth, etc., terms, every succeeding term is very small compared with the preceding. Bearing in mind that

$$\int_{-\infty}^{+\infty}\phi(\varDelta)d\varDelta = \mathrm{I},$$

and putting

$$\frac{\mathrm{I}}{\tau}\int_{-\infty}^{+\infty}\frac{\varDelta^2}{2}\phi(\varDelta)d\varDelta = D,$$

and taking into consideration only the first and third terms on the right-hand side, we get from this equation

(3) $$\frac{\partial f}{\partial t} = D\frac{\partial^2 f}{\partial x^2}.$$

This is the well-known differential equation for diffusion, and we recognise that D is the coefficient of diffusion.

Another important consideration can be related to this method of development. We have assumed that the single particles are all referred to the same co-ordinate system. But this is unnecessary, since the movements of the single particles are mutually independent. We will now refer the motion of each particle to a co-ordinate system whose origin coincides at the time $t = 0$ with the position of the centre of gravity of the particles in question ; with this difference, that $f(x, t)dx$ now gives the number of the particles whose x co-ordinate has increased between the time $t = 0$ and the time $t = t$, by a quantity which lies between x and $x + dx$. In this case also the function f must satisfy, in its changes, the equation (1). Further, we must evidently have for $x \gtrless 0$ and $t = 0$,

$$f(x, t) = 0 \text{ and } \int_{-\infty}^{+\infty} f(x, t)dx = n.$$

The problem, which accords with the problem of the diffusion outwards from a point (ignoring possibilities of exchange between the diffusing particles) is now mathematically completely defined (9) ; the solution is

$$f(x, t) = \frac{n}{\sqrt{4\pi D}} \frac{e^{-\frac{x^2}{4Dt}}}{\sqrt{t}} \qquad . \qquad . \quad (10)$$

The probable distribution of the resulting displacements in a given time t is therefore the same as that of fortuitous error, which was to be expected. But it is significant how the constants in the exponential term are related to the coefficient of diffusion. We will now calculate with the help of this equation the displacement λ_x in the direction of the X-axis which a particle experiences on an average, or—more accurately expressed—the square root of the arithmetic mean of the squares of displacements in the direction of the X-axis; it is

$$\lambda_x = \sqrt{\overline{x^2}} = \sqrt{2Dt} \qquad . \qquad . \quad (11)$$

The mean displacement is therefore proportional to the square root of the time. It can easily be shown that the square root of the mean of the squares of the total displacements of the particles has the value $\lambda_x\sqrt{3}$ . . . (12)

§ 5. Formula for the Mean Displacement of Suspended Particles. A New Method of Determining the Real Size of the Atom

In § 3 we found for the coefficient of diffusion D of a material suspended in a liquid in the form of small spheres of radius P—

$$D = \frac{RT}{N} \cdot \frac{1}{6\pi kP}.$$

Further, we found in § 4 for the mean value of the displacement of the particles in the direction of the X-axis in time t—

$$\lambda_x = \sqrt{2Dt}.$$

By eliminating D we obtain

$$\lambda_x = \sqrt{t} \cdot \sqrt{\frac{RT}{N} \frac{1}{3\pi kP}}.$$

This equation shows how λ_x depends on T, k, and P.

We will calculate how great λ_x is for one second, if N is taken equal to $6 \cdot 10^{23}$ in accordance with the kinetic theory of gases, water at $17°$ C. is chosen as the liquid ($k = 1 \cdot 35 \cdot 10^{-2}$), and the diameter of the particles ·001 mm. We get

$$\lambda_x = 8 \cdot 10^{-5} \text{ cm.} = 0 \cdot 8\mu.$$

The mean displacement in one minute would be, therefore, about 6μ.

On the other hand, the relation found can be used for the determination of N. We obtain

$$N = \frac{1}{\lambda_x^2} \cdot \frac{RT}{3\pi kP}.$$

It is to be hoped that some enquirer may succeed shortly in solving the problem suggested here, which is so important in connection with the theory of Heat. (13)

Berne, *May*, 1905.

(Received, 11 *May*, 1905.)

II

ON THE THEORY OF THE BROWNIAN MOVEMENT

(From the *Annalen der Physik* (4), **19**, 1906, pp. 371-381)

SOON after the appearance of my paper (*) on the movements of particles suspended in liquids demanded by the molecular theory of heat, Siedentopf (of Jena) informed me that he and other physicists—in the first instance, Prof. Gouy (of Lyons)—had been convinced by direct observation that the so-called Brownian motion is caused by the irregular thermal movements of the molecules of the liquid.(†)

Not only the qualitative properties of the Brownian motion, but also the order of magnitude of the paths described by the particles correspond completely with the results of the theory. I will not attempt here a comparison of the slender experimental material at my disposal with the

(*) A. Einstein, *Ann. d. Phys.*, **17**, p. 549, 1905.

(†) M. Gouy, *Journ. de Phys.* (2), **7**, 561, 1888.

results of the theory, but will leave this comparison to those who may be handling the experimental side of the problem.

The following paper will amplify in some points the author's own paper mentioned above. We will derive here not only the translational movement, but also the rotational movement of suspended particles, for the simplest special case where the particles have a spherical form. We will show further, up to how short a time of observation the results given in that discussion hold true.

To derive these we will use here a more general method, partly in order to show how the Brownian motion is related to the fundamentals of the molecular theory of heat, partly to be able to develop the formula for the translational and the rotational movement in a single discussion. Suppose, accordingly, that α is a measurable parameter of a physical system in thermal equilibrium, and assume that the system is in the so-called neutral equilibrium for every (possible) value of α. According to classical thermodynamics, which differentiates in principle between heat and other kinds of energy, spontaneous alterations of α cannot occur ; according to the molecular theory of heat, it is otherwise. In the following we will investigate according to what laws the alterations implied by the latter theory take place. We must then apply these laws to the following special cases :—

1. α is the X-co-ordinate of the centre of gravity of a suspended particle of spherical form in a homogeneous liquid (not subject to gravitation).

2. α is the angle which determines the position of a particle, rotatable about a diameter, that is suspended in a liquid.

§ 1. ON A CASE OF THERMODYNAMIC EQUILIBRIUM

Suppose a physical system placed in an environment of absolute temperature T, which system has thermal interchange with the environment and is in a state of thermal equilibrium. This system (which therefore has also the absolute temperature T) is fully defined in the terms of the molecular theory of heat (*) by the variables of condition $p_1 \ldots p_n$. In the special cases to be considered, the co-ordinates and velocity-components of all the atoms forming the particular system can be chosen as the variables of condition $p_1 \ldots p_n$.

(*) Cf. *Ann. d. Phys.*, **11**, p. 170, 1903 ; **17**, p. 549, 1905.

For the probability, that at any arbitrarily-chosen moment of time the variables of condition $p_1 \ldots p_n$ lie within an indefinitely small n-fold domain $(dp_1 \ldots dp_n)$, the equation holds— (*)

(1) $$dw = Ce^{-\frac{N}{RT}E}dp_1 \ldots dp_n,$$

where C is a constant, R the universal constant of the gas equation, N the number of the actual molecules in a gram-molecule, and E the energy.

Suppose α is a parameter of the system that can be measured, and suppose each set of values $p_1 \ldots p_n$ implies a definite value α, we will indicate by $A d\alpha$ the probability that at any arbitrarily-chosen moment of time the value of the parameter α lies between α and $\alpha + d\alpha$. Then

(2) $$A d\alpha = \int_{d\alpha} Ce^{-\frac{N}{RT}E} dp_1 \ldots dp_n,$$

where the integration is taken over all combinations of values of the variables of condition, whose α value lies between α and $\alpha + d\alpha$.

We will confine ourselves to the case, which is clear without further discussion from the nature of the problem, where all (possible) values of α have the same probability (frequency) ; where, therefore, the quantity A is independent of α.

(*) L.c. §§ 3 and 4.

A second physical system can now be studied which differs from that already considered only in that a force, of potential $\Phi(\alpha)$, dependent solely on α, is acting on the system. If E is the energy of the former system, then the energy of the present system will be $E + \Phi$, so that we obtain a relation analogous to the equation (1)—

$$dw' = C'e^{-\frac{N}{RT}(E + \Phi)}dp_1 \ldots dp_n.$$

From this can be deduced an expression analogous to the equation (2), for the probability dW that at any arbitrarily-chosen moment of time the value of α lies between α and $\alpha + d\alpha$—

$$(I) \quad \begin{cases} dW = \int C'e^{-\frac{N}{RT}(E + \Phi)}dp_1 \ldots dp_n \\ \qquad\qquad = \frac{C'}{C}e^{-\frac{N}{RT}\Phi}A d\alpha = A'e^{-\frac{N}{RT}}d\alpha \end{cases}$$

where A' is independent of α.

This relation, which corresponds exactly with the exponential law frequently used by Boltzmann (14) in his investigations in the theory of gases, is characteristic of the molecular theory of heat. It explains how far a parameter of a system, when a constant external force is applied, can diverge from the value which corresponds to stable equilibrium, as the result of irregular molecular movements.

§ 2. Examples of the Application of the Equation obtained in § 1

We will consider a body whose centre of gravity can move along a straight line (the X-axis of a co-ordinate system). The body is surrounded by a gas and there is thermal and mechanical equilibrium. According to the molecular theory, as the result of the irregularity of the impacts of the molecules, the body will move backwards and forwards along the straight line in an irregular manner, so that in this movement preference is given to no particular point in the straight line—provided that no forces act on the body in the direction of the straight line, other than the forces of impact of the molecules. The abscissa x of the centre of gravity is then a parameter of the system, which possesses the properties indicated above for the parameter α.

We will introduce now a force $K = -Mx$ acting on the body in the direction of the straight line. Then, according to the molecular theory, the centre of gravity of the body will also execute irregular movements, without departing far from the point $x = 0$, while according to classical thermodynamics it must remain stationary at the point $x = 0$. According to the molecular theory (formula (I)),

$$dW = A'e^{-\frac{N}{RT}M\frac{x^2}{2}}dx,$$

is equal to the probability that at an arbitrarily-chosen moment of time the value of the abscissa x lies between x and $x + dx$. From this we find the mean distance of the centre of gravity from the point $x = 0$—

$$\sqrt{\overline{x^2}} = \frac{\int_{-\infty}^{+\infty} x^2 A'e^{-\frac{N}{RT}\frac{Mx^2}{2}}dx}{\int_{-\infty}^{+\infty} A'e^{-\frac{N}{RT}\frac{Mx^2}{2}}dx} = \sqrt{\frac{RT}{NM}}.$$

In order that $\sqrt{\overline{x^2}}$ may be large enough to be capable of direct observation, the force establishing the equilibrium position of the body must be very small. Let us put for the lower limit of observation $\sqrt{\overline{x^2}} = 10^{-4}$ cms. ; then, if $T = 300$ we get $M = 5 \cdot 10^{-6}$ approximately. In order that the body may carry out vibrations visible in the microscope the force acting on it when the displacement is 1 cm. must not exceed five millionths of a dyne. (15)

We will add a further theoretical observation to the equation we have obtained. Suppose the body in question carries an electrical charge distributed over a very small space, and that the gas surrounding the body is so tenuous that the body carries out vibrations of only slightly modified sine-form in the surrounding gas. Then the body will radiate electric waves into the space, and will absorb energy from the radiation of the surrounding space ; it brings about, therefore, an energy exchange between the radiation and the gas. We can derive the limiting law for temperature-radiation, which appears to hold for long wave-lengths and for high temperatures, if we lay down the condition that the body in question emits on the average just as much radiation as it absorbs. We obtain thus (*) the following formula for the density of radiation corresponding to the frequency ν—

$$\rho\nu = \frac{R}{N}\frac{8\pi\nu^2}{L^3}T,$$

where L is the velocity of light. (16)

The radiation formula of Planck (†) can be transformed into this expression when the frequency is small and the temperature is high. The quantity N can be determined from the coefficients in the limiting law, and we obtain thus Planck's calculation of the elementary quanta. The fact that we obtain in the manner indicated not the true law of radiation, but only a limiting law,

(*) Cf. *Ann. d. Phys.*, **17**, p. 132, 1905 ; §§ 1 and 2.
(†) M. Planck, *Ann. d. Phys.*, **1**, p. 99, 1900.

appears to me to have an explanation in a fundamental incompleteness in our physical conceptions.

We will now apply the formula (I) to determine
how small a suspended particle must be in order
that it may remain permanently suspended in
spite of gravitation. We can confine ourselves
to the case where the particle is of greater density
than the liquid, since the opposite case is fully
analogous. If v is the volume of the particle,
ρ its density, ρ_0 the density of the liquid, g the
acceleration of gravity, and x the vertical distance
of a point from the bottom of the vessel, equation
(I) gives

$$dW = \text{const.}\ e^{-\frac{N}{RT}v(\rho-\rho_0)g\,dx}dx \qquad (17)$$

We shall find, therefore, that suspended particles
are able to remain suspended when for values of
x which do not escape observation on account of
their minuteness, the quantity

$$\frac{RT}{N}v(\rho-\rho_0)gx$$

has not too high a value—with the understanding
that particles which may reach the bottom of the
vessel are not held fast there by some peculiar
condition of the latter.

§ 3. On the Changes in the Parameter α Brought about by Thermal Motion

We will return to the general case considered in § 1, for which we have derived equation (I). However, for the sake of a simpler mode of expression and presentation, we will now assume that there are a very large number (n) of identical systems of the type indicated there ; we have, then, to do with numbers in place of probabilities. Equation (I) then expresses :—

Of N systems, in

(Ia) $$dn = \phi e^{-\frac{N}{RT}\Phi} d\alpha = F(\alpha)d\alpha$$

the value of the parameter α at an arbitrarily-chosen moment of time falls between α and α + $d\alpha$.

We will use this relation to ascertain the magnitude of the irregular changes of the parameter α produced by the irregular thermal phenomena. For this purpose we express in symbols that the function $F(\alpha)$ does not alter within the time-interval t under the combined effect of the force corresponding to the potential Φ and of the irregular thermal processes ; t indicates here so small a time that the corresponding changes of the quantity α of the single systems can be looked upon as indefinitely small changes in the argument of the function $F(\alpha)$.

If lengths are marked out from a definite zero point along a straight line, each numerically equal to the quantities α, each system determines a point (α) on this straight line. Now, during the time t precisely as many points (determined each by a system) must pass through a particular point (α_0) in one direction, as in the other direction. The force corresponding to the potential Φ produces a change in α of the magnitude

$$\Delta_1 = - B\frac{\partial}{\partial\alpha}t,$$

where B is independent of α, that is, the velocity of change of α is proportional to the imposed force and independent of the value of the parameter. We will call the factor B the "Mobility of the system in respect to α."

If, therefore, the external force operates, whilst the quantity α is not changed by the irregular molecular thermal processes, there will pass through the point (α_0) during the time t

$$n_1 = B\left(\frac{\partial\Phi}{\partial\alpha}\right)_{\alpha\,=\,\alpha_0} \cdot tF(\alpha_0)$$

points (determined each by a system) in the direction of the negative side.

Suppose, further, that the probability that the parameter α of a system experiences a change in the time t (on account of the irregular thermal processes) whose value lies between Δ and $\Delta + d\Delta$, is equal to $\psi(\Delta)$, where $\psi(\Delta) = \psi(-\Delta)$ and ψ is independent of α.

The number of points (each determined by a system) which pass through the point (α_0) during the time t in the direction of the positive side as the result of the irregular thermal processes is then

$$n_2 = \int_{\Delta = 0}^{\Delta = \infty} F(\alpha_0 - \Delta)\chi(\Delta)d\Delta,$$

where we put

$$\int_{\Delta}^{\infty}\psi(\Delta)d\Delta = \chi(\Delta).$$

The number of points which pass in the direction of the negative side as the result of the irregular thermal processes is

$$n_3 = \int_{\Delta}^{\infty} F(\alpha_0 + \Delta)\chi(\Delta)d\Delta.$$

The mathematical expression for the invariability of the function F is therefore

$$-n_1 + n_2 - n_3 = 0.$$

If we introduce the expressions found for n_1, n_2, n_3, and bear in mind that Δ is indefinitely small, or that $\psi(\Delta)$ only differs from zero for indefinitely small values of Δ, respectively, we obtain after simple manipulation

$$B\left(\frac{\partial\Phi}{\partial\alpha}\right)_{\alpha=\alpha_0} F(\alpha_0)t + \tfrac{1}{2}F'(\alpha_0)\overline{\Delta^2} = 0 \quad . \quad (18)$$

Here

$$\overline{\varDelta^2} = \int_{-\infty}^{+\infty} \varDelta^2 \psi(\varDelta) d\varDelta$$

signifies the mean value of the squares of the changes in the quantities α produced by the irregular thermal processes during the time t. From this relation we obtain, with due reference, the equation (Ia)—

(II) $$\sqrt{\overline{\varDelta^2}} = \sqrt{\frac{2R}{N}} \cdot \sqrt{BTt}.$$

Here R is the constant of the gas-equation ($8\cdot31 \cdot 10^7$), N the number of the actual molecules in a gram molecule (about $6\cdot10^{23}$) (19), B the " mobility of the system in respect to the parameter α," T the absolute temperature, t the time within which the changes in α take place that are produced by the irregular thermal processes.

§ 4. APPLICATION OF THE EQUATION DERIVED, TO THE BROWNIAN MOTION

We will now calculate with the help of equation (II), in the first instance, the mean displacement which a body of spherical form suspended in a liquid experiences in the time t in a definite direction (the direction of the X-axis in a co-ordinate system). For this purpose we must insert the corresponding value for B in the former equation.

If a force K acts on a sphere of radius P, which is suspended in a liquid of viscosity k, it will move (*) with a velocity $K/6\pi kP$. (6) Accordingly we can put

$$B = \frac{1}{6\pi kP},$$

so that we get—in conformity with the paper mentioned above—for the mean displacement of the suspended sphere in the direction of the X-axis the value

$$\sqrt{\overline{\Delta_x^2}} = \sqrt{t}\sqrt{\frac{RT}{N}\frac{1}{3\pi kP}}.$$

Secondly, we will consider the case where the sphere in question is mounted in the liquid so as to be freely rotatable, without friction, about its diameter, and investigate the mean rotation $\sqrt{\overline{\Delta^2}}$ of the sphere during the time t, as the result of the irregular thermal processes.

If the moment D acts on a sphere of radius P, which is mounted so as to be capable of rotation in a liquid of viscosity k, it rotates with the angular velocity (†)

$$\psi = \frac{D}{8\pi kP^3}.$$

(*) Cf. G. Kirchhoff, " Lectures on Mechanics," Lect. 26.

(†) *Ibid.*

We have, therefore, to put

$$B = \frac{1}{8\pi k P^3}.$$

Accordingly, we get

$$\sqrt{\overline{\varDelta_r^2}} = \sqrt{t}\sqrt{\frac{RT}{N}\frac{1}{4\pi k P^3}} \qquad . \qquad (20)$$

The angular motion produced by the molecular motion decreases therefore with increasing P much more rapidly than the progressive motion.

For $P = 0\cdot 5$ mm. and water at $17°$ the formula gives for the angle described on an average in one second about 11 seconds of arc ; in an hour about 11 minutes of arc. For $P = 0\cdot 5\mu$ and water at $17°$ we get for $t = 1$ second about $100°$ of arc.

In the case of a totally unconstrained suspended particle, three mutually independent angular motions of this kind are possible.

The formula derived for $\sqrt{\overline{\varDelta^2}}$ can be applied further to other cases. For example, if for B is inserted the reciprocal of the electrical resistance of a closed circuit, the formula states how much electricity will flow on an average during the time t across any particular cross-section of the conductor, which relation again is connected with the limiting law for the radiation of a black body for long wave-lengths and high temperatures. (21)

However, since I have been able to find no further consequences that can be checked up experimentally, it appears to me to be unprofitable to consider other special cases.

§ 5. On the Limits of Application of the Formula for $\sqrt{\overline{\Delta^2}}$

It is clear that the formula (II) cannot be applied for any arbitrarily small time. For the mean velocity of change of α as the result of the thermal processes

$$\frac{\sqrt{\overline{\Delta^2}}}{t} = \sqrt{\frac{2RTB}{N}} \cdot \frac{1}{\sqrt{t}}$$

becomes infinitely great for an indefinitely small interval of time t; which is evidently impossible, since in that case each suspended particle would move with an infinitely great instantaneous velocity. The reason is that we have implicitly assumed in our development that the events during the time t are to be looked upon as phenomena independent of the events in the time immediately preceding. But this assumption becomes harder to justify the smaller the time t is chosen.

If the instantaneous value of the velocity of change, at a time $z = 0$, is

$$\frac{d\alpha}{dt} = \beta_0,$$

and if the velocity of change β is not affected by the irregular thermal processes in a certain subsequent interval of time, but the change of β is solely determined by the passive resistance $(1/B)$, then this relation will hold for $d\beta/dz$:—

$$- \mu \frac{d\beta}{dz} = \frac{\beta}{B}.$$

μ is here defined by the condition that $\mu(\beta^2/2)$ must be the energy corresponding to the velocity of change β. In the case, therefore, of translational movement of the sphere $\mu(\beta^2/2)$ is, for example, the kinetic energy of the liquid carried with it. It follows by integration

$$\beta = \beta_0 e^{-\frac{z}{\mu B}}.$$

We conclude from this result that the formula (II) only holds for intervals of time which are large compared with μB. (22)

For small bodies of 1μ diameter and unit density in water at room-temperature, the lower limit of availability of the formula (II) is about 10^{-7} seconds ; this lower limit for the interval of time increases in proportion to the square of the radius of the body. Both hold for the translational as well as for the rotational motion of the particle.

Berne, *December*, 1905.

(Received, 19 *December*, 1905.)

III

A NEW DETERMINATION OF MOLECULAR DIMENSIONS

(From the *Annalen der Physik* (4), **19**, 1906, pp. 289-306. Corrections, *ibid.*, **34**, 1911, pp. 591-592.) (23)

THE kinetic theory of gases made possible the earliest determinations of the actual dimensions of the molecules, whilst physical phenomena observable in liquids have not, up to the present, served for the calculation of molecular dimensions. The explanation of this doubtless lies in the difficulties, hitherto unsurpassable, which discourage the development of a molecular kinetic theory of liquids that will extend to details. It will be shown now in this paper that the size of the molecules of the solute in an undissociated dilute solution can be found from the viscosity of the solution and of the pure solvent, and from the rate of diffusion of the solute into the solvent, if the volume of a molecule of the solute is large compared with the volume of a molecule of the solvent. For such a solute molecule will behave approximately, with respect to its mobility in the solvent, and in respect to its influence on the viscosity of the latter, as a solid body suspended in the solvent, and it will be allowable to apply to the motion of the solvent in the immediate neighbourhood of a molecule the hydrodynamic equations, in which the liquid is considered homogeneous, and, accordingly, its molecular structure is ignored. We will choose for the shape of the solid bodies, which shall represent the solute molecules, the spherical form.

§ 1. On the Effect on the Motion of a Liquid of a Very Small Sphere Suspended in it

As the subject of our discussion, let us take an incompressible homogeneous liquid with viscosity k, whose velocity-components u, v, w will be given as functions of the co-ordinates x, y, z, and of the time. Taking an arbitrary point x_0, y_0, z_0, we will imagine that the functions u, v, w are developed according to Taylor's theorem as functions of $x - x_0$, $y - y_0$, $z - z_0$, and that a domain G is marked out around this point so small that within it only the linear terms in this expansion have to be considered.   The motion of the liquid contained in G can then be looked upon in the familiar manner as the result of the superposition of three motions, namely,

1. A parallel displacement of all the particles of the liquid without change of their relative position.

2. A rotation of the liquid without change of the relative position of the particles of the liquid.

3. A movement of dilatation in three directions at right angles to one another (the principal axes of dilatation).

We will imagine now a spherical rigid body in the domain G, whose centre lies at the point x_0, y_0, z_0, and whose dimensions are very small compared with those of the domain G. We will further assume that the motion under consideration is so slow that the kinetic energy of the sphere is negligible as well as that of the liquid. It will be further assumed that the velocity components of an element of surface of the sphere show agreement with the corresponding velocity components of the particles of the liquid in the immediate neighbourhood, that is, that the contact-layer (thought of as continuous) also exhibits everywhere a viscosity-coefficient that is not vanishingly small.

It is clear without further discussion that the sphere simply shares in the partial motions 1 and 2, without modifying the motion of the neighbouring liquid, since the liquid moves as a rigid body in these partial motions ; and that we have ignored the effects of inertia.

But the motion 3 will be modified by the presence of the sphere, and our next problem will be to investigate the influence of the sphere on this motion of the liquid. We will further refer the motion 3 to a co-ordinate system whose axes are parallel to the principal axes of dilatation, and we will put

$$x - x_0 = \xi,$$
$$y - y_0 = \eta,$$
$$z - z_0 = \zeta,$$

then the motion can be expressed by the equations

(1)
$$u_0 = A\xi,$$
$$v_0 = B\eta,$$
$$w_0 = C\zeta,$$

in the case when the sphere is not present. A, B, C are constants which, on account of the incompressibility of the liquid, must fulfil the condition

(2) $$A + B + C = 0 \qquad . \qquad . \quad (24)$$

Now, if the rigid sphere with radius P is introduced at the point x_0, y_0, z_0, the motions of the liquid in its neighbourhood are modified. In the following discussion we will, for the sake of convenience, speak of P as " finite " ; whilst the values of ξ, η, ζ, for which the motions of the liquid are no longer appreciably influenced by the sphere, we will speak of as " infinitely great."

Firstly, it is clear from the symmetry of the motions of the liquid under consideration that there can be neither a translation nor a rotation of the sphere accompanying the motion in question, and we obtain the limiting conditions

$$u = v = w = 0 \text{ when } \rho = P$$

where we have put

$$\rho = \sqrt{\xi^2 + \eta^2 + \zeta^2} > 0.$$

Here u, v, w are the velocity-components of the motion now under consideration (modified by the sphere). If we put

$$(3) \qquad \begin{aligned} u &= A\xi + u_1, \\ v &= B\eta + v_1, \\ w &= C\zeta + w_1, \end{aligned}$$

since the motion defined by equation (3) must be transformed into that defined by equations (1) in the "infinite" region, the velocities u_1, v_1, w_1 will vanish in the latter region.

The functions u, v, w must satisfy the hydro-dynamic equations with due reference to the viscosity, and ignoring inertia. Accordingly, the following equations will hold :— (*)

$$(4) \quad \left\{ \frac{\partial p}{\partial \xi} = k\Delta u, \; \frac{\partial p}{\partial \eta} = k\Delta v, \; \frac{\partial p}{\partial \zeta} = k\Delta w, \; \frac{\partial u}{\partial \xi} + \frac{\partial v}{\partial \eta} + \frac{\partial w}{\partial \zeta} = 0, \right.$$

where Δ stands for the operator

$$\frac{\partial^2}{\partial \xi^2} + \frac{\partial^2}{\partial \eta^2} + \frac{\partial^2}{\partial \zeta^2}$$

and p for the hydrostatic pressure.

Since the equations (1) are solutions of the equations (4) and the latter are linear, according to (3) the quantities u_1, v_1, w_1 must also satisfy the equations (4). I have determined u_1, v_1, w_1, and p, according to a method given in the lecture of Kirchhoff quoted in § 4 (†), and find

(*) G. Kirchhoff, " Lectures on Mechanics," Lect. 26.

(†) " From the equations (4) it follows that $\Delta p = 0$. If p is chosen in accordance with this condition, and a function V is determined which satisfies the equation

$$\Delta V = \frac{1}{k}p,$$

then the equations (4) are satisfied if we put

$$u = \frac{\partial V}{\partial \xi} + u', \quad v = \frac{\partial V}{\partial \eta} + v', \quad w = \frac{\partial V}{\partial \zeta} + w'$$

and chose u', v', w', so that $\Delta u' = 0$, $\Delta v' = 0$, and $\Delta w' = 0$, and

$$\frac{\partial u'}{\partial \xi} + \frac{\partial v'}{\partial \eta} + \frac{\partial w'}{\partial \zeta} = -\frac{1}{k}p."$$

$$p = -\frac{5}{3}kP^3\left\{A\frac{\partial^2\left(\frac{1}{\rho}\right)}{\partial\xi^2} + B\frac{\partial^2\left(\frac{1}{\rho}\right)}{\partial\eta^2}\right.$$

$$\left. + C\frac{\partial^2\left(\frac{1}{\rho}\right)}{\partial\zeta^2}\right\} + \text{const.}$$

$$(5)\begin{cases} u = A\xi - \frac{5}{3}P^3A\frac{\xi}{\rho^3} - \frac{\partial D}{\partial\xi}, \\[2mm] v = B\eta - \frac{5}{3}P^3B\frac{\eta}{\rho^3} - \frac{\partial D}{\partial\eta}, \\[2mm] w = C\zeta - \frac{5}{3}P^3C\frac{\zeta}{\rho^3} - \frac{\partial D}{\partial\zeta}, \end{cases}$$

Now if we put

$$\frac{p}{k} = 2c\frac{\partial^2\frac{1}{\rho}}{\partial\xi^2}$$

and in agreement with this

$$V = c\frac{\partial^2\rho}{\partial\xi^2} + b\frac{\partial^2\frac{1}{\rho}}{\partial\xi^2} + \frac{a}{2}\left(\xi^2 - \frac{\eta^2}{2} - \frac{\zeta^2}{2}\right)$$

and

$$u' = -2c\frac{\partial\frac{1}{\rho}}{\partial\xi}, \; v' = 0, \; w' = 0,$$

the constants a, b, c can be chosen so that when $\rho = p$, $u = v = w = 0$. By superposition of three similar solutions we obtain the solution given in the equations (5) and (5a).

where

(5a)
$$
\left\{
\begin{aligned}
D = & A\left\{\frac{5}{6}P^3\frac{\partial^2\rho}{\partial\xi^2} + \frac{1}{6}P^5\frac{\partial^2\left(\frac{1}{\rho}\right)}{\partial\xi^2}\right\} \\
& + B\left\{\frac{5}{6}P^3\frac{\partial^2\rho}{\partial\eta^2} + \frac{1}{6}P^5\frac{\partial^2\left(\frac{1}{\rho}\right)}{\partial\eta^2}\right\} \\
& + C\left\{\frac{5}{6}P^3\frac{\partial^2\rho}{\partial\zeta^2} + \frac{1}{6}P^5\frac{\partial^2\left(\frac{1}{\rho}\right)}{\partial\zeta^2}\right\}.
\end{aligned}
\right.
$$

It is easy to see that the equations (5) are solutions of the equations (4). Then, since

$$
\Delta\xi = 0, \quad \Delta\frac{1}{\rho} = 0, \quad \Delta\rho = \frac{2}{\rho}
$$

and

$$
\Delta\left(\frac{\xi}{\rho^3}\right) = -\frac{\partial}{\partial\xi}\left\{\Delta\left(\frac{1}{\rho}\right)\right\} = 0,
$$

we get

$$
k\Delta u = -k\frac{\partial}{\partial\xi}\{\Delta D\}
$$

$$
= -k\frac{\partial}{\partial\xi}\left\{\frac{5}{3}P^3A\frac{\partial^2\frac{1}{\rho}}{\partial\xi^2} + \frac{5}{3}P^3B\frac{\partial^2\frac{1}{\rho}}{\partial\eta^2} + \cdots\right\}.
$$

But the last expression obtained is, according to the first of the equations (5), identical with $dp/d\xi$. In similar manner, we can show that the second and third of the equations (4) are satisfied. We obtain further—

$$\frac{\partial u}{\partial \xi} + \frac{\partial v}{\partial \eta} + \frac{\partial w}{\partial \zeta} = (A + B + C)$$

$$+ \frac{5}{3}P^3\left\{A\frac{\partial^2\left(\frac{1}{\rho}\right)}{\partial \xi^2} + B\frac{\partial^2\left(\frac{1}{\rho}\right)}{\partial \eta^2} + C\frac{\partial^2\left(\frac{1}{\rho}\right)}{\partial \zeta^2}\right\} - \Delta D.$$

But since, according to equation (5a),

$$\Delta D = \frac{5}{3}P^3\left\{A\frac{\partial^2\left(\frac{1}{\rho}\right)}{\partial \xi^2} + B\frac{\partial^2\left(\frac{1}{\rho}\right)}{\partial \eta^2} + C\frac{\partial^2\left(\frac{1}{\rho}\right)}{\partial \zeta^2}\right\},$$

it follows that the last of the equations (4) is satisfied. As for the boundary conditions, our equations for u, v, w are transformed into the equations (1) only when ρ is indefinitely large. By inserting the value of D from the equation (5a) in the second of the equations (5) we get

$$(6) \quad u = A\xi - \frac{5}{2}\frac{P^3}{\rho^5}\xi(A\xi^2 + B\eta^2 + C\zeta^2)$$

$$+ \frac{5}{2}\frac{P^5}{\rho^7}\xi(A\xi^2 + B\eta^2 + C\zeta^2) - \frac{P^5}{\rho^5}A\xi \quad (25)$$

We know that u vanishes when $\rho = P$. On the grounds of symmetry the same holds for v and w. We have now demonstrated that in the equations (5) a solution has been obtained to satisfy both the equations (4) and the boundary conditions of the problem.

It can also be shown that the equations (5) are the only solutions of the equations (4) consistent with the boundary conditions of the problem. The proof will only be indicated here. Suppose that, in a finite space, the velocity-components of a liquid u, v, w satisfy the equations (4). Now, if another solution U, V, W of the equations (4) can exist, in which on the boundaries of the sphere in question $U = u$, $V = v$, $W = w$, then $(U - u, V - v, W - w)$ will be a solution of the equations (4), in which the velocity-components vanish at the boundaries of the space. Accordingly, no mechanical work can be done on the liquid contained in the space in question. Since we have ignored the kinetic energy of the liquid, it follows that the work transformed into heat in the space in question is likewise equal to zero. Hence we infer that in the whole space we must have $u = u'$, $v = v'$, $w = w'$, if the space is bounded, at least in part, by stationary walls. By crossing the boundaries, this result can also be extended to the case when the space in question is infinite, as in the case considered above. We can show thus that the solution obtained above is the sole solution of the problem.

We will now place around the point x_0, y_0, z_0 a sphere of radius R, where R is indefinitely large compared with P, and will calculate the energy which is transformed into heat (per unit of time) in the liquid lying within the sphere. This energy W is equal to the mechanical work done on the liquid. If we call the components of the pressure exerted on the surface of the sphere of radius R, X_n, Y_n, Z_n, then

$$W = \int (X_n u + V_n v + Z_n w) ds$$

where the integration is extended over the surface of the sphere of radius R.

Here

$$X_n = - \left(X_\xi \frac{\xi}{\rho} + X_\eta \frac{\eta}{\rho} + X_\zeta \frac{\zeta}{\rho} \right),$$

$$Y_n = - \left(Y_\xi \frac{\xi}{\rho} + Y_\eta \frac{\eta}{\rho} + Y_\zeta \frac{\zeta}{\rho} \right),$$

$$Z_n = - \left(Z_\xi \frac{\xi}{\rho} + Z_\eta \frac{\eta}{\rho} + Z_\zeta \frac{\zeta}{\rho} \right),$$

where

$$X_\xi = p - 2k\frac{\partial u}{\partial \xi}, \qquad Y_\zeta = Z_\eta = -k\left(\frac{\partial v}{\partial \zeta} + \frac{\partial w}{\partial \eta}\right)$$

$$Y_\eta = p - 2k\frac{\partial v}{\partial \eta}, \qquad Z_\xi = X_\zeta = -k\left(\frac{\partial w}{\partial \xi} + \frac{\partial u}{\partial \zeta}\right)$$

$$Z_\zeta = p - 2k\frac{\partial w}{\partial \zeta}, \qquad X_\eta = Y_\xi = -k\left(\frac{\partial u}{\partial \eta} + \frac{\partial v}{\partial \xi}\right).$$

The expressions for u, v, w are simplified when we note that for $\rho = R$ the terms with the factor P^5/ρ^5 vanish.

We have to put

$$u = A\xi - \frac{5}{2}P^3\frac{\xi(A\xi^2 + B\eta^2 + C\zeta^2)}{\rho^5}$$

(6a) $$v = B\eta - \frac{5}{2}P^3\frac{\eta(A\xi^2 + B\eta^2 + C\zeta^2)}{\rho^5}$$

$$w = C\zeta - \frac{5}{2}P^3\frac{\zeta(A\xi^2 + B\eta^2 + C\zeta^2)}{\rho^5}$$

For p we obtain from the first of the equations (5) by corresponding omissions

$$p = -5kP^3\frac{A\xi^2 + B\eta^2 + C\zeta^2}{\rho^5} + \text{const.}$$

We obtain first

$$X_\xi = -2kA + 10kP^3\frac{A\xi^2}{\rho^5} - 25kP^3\frac{\xi^2(A\xi^2 + B\eta^2 + C\zeta^2)}{\rho^7}$$

$$X_\eta = +5kP^3\frac{(A+B)\xi\eta}{\rho^5} - 25kP^3\frac{\xi\eta(A\xi^2 + B\eta^2 + C\zeta^2)}{\rho^7} \qquad (23)$$

$$X_\zeta = +5kP^3\frac{(A+C)\xi\zeta}{\rho^5} - 25kP^3\frac{\xi\zeta(A\xi^2 + B\eta^2 + C\zeta^2)}{\rho^7}$$

and from this

$$X_n = 2Ak\frac{\xi}{\rho} - 5AkP^3\frac{\xi}{\rho^4} + 20kP^3\frac{\xi(A\xi^2 + B\eta^2 + C\zeta^2)}{\rho^6}. \quad (23)$$

With the aid of the expressions for Y_n and Z_n, obtained by cyclic exchange, we get, ignoring all terms which involve the ratio P/ρ raised to any power higher than the third,

$$X_n u + Y_n v + Z_n w = \frac{2k}{\rho}(A^2\xi^2 + B^2\eta^2 + C^2\zeta^2)$$

$$-5k\frac{P^3}{\rho^4}(A^2\xi^2 + B^2\eta^2 + C^2\zeta^2) + 15k\frac{P^3}{\rho^6}(A\xi^2 + B\eta^2 + C\zeta^2)^2. \quad (23)$$

If we integrate over the sphere and bear in mind that

$\int ds = 4R^2\pi,$

$\int \xi^2 ds = \int \eta^2 ds = \int \zeta^2 ds = \frac{4}{3}\pi R^4,$

$\int \xi^4 ds = \int \eta^4 ds = \int \zeta^4 ds = \frac{4}{5}\pi R^6,$

$\int \eta^2\zeta^2 ds = \int \zeta^2\xi^2 ds = \int \xi^2\eta^2 ds = \frac{4}{15}\pi R^6,$

$\int (A\xi^2 + B\eta^2 + C\zeta^2)^2 ds = \frac{8}{15}\pi R^6(A^2 + B^2 + C^2), \quad (23)$

we obtain

(7) $\quad W = \frac{8}{3}\pi R^3 k\delta^2 + \frac{4}{3}\pi P^3 k\delta^2 = 2\delta^2 k\left(V + \frac{\Phi}{2}\right), \quad (23)$

where we put

$$\delta^2 = A^2 + B^2 + C^2,$$

$$\frac{4\pi}{3}R^3 = V \text{ and } \frac{4}{3}\pi P^3 = \Phi$$

If the suspended sphere were not present ($\Phi = 0$), then we should get for the energy used up in the volume V,

(7a) $\qquad\qquad W = 2\delta^2 kV.$

On account of the presence of the sphere, the energy used up is therefore diminished by $\delta^2 k\Phi$.

(26)

§ 2. Calculation of the Viscosity-Coefficient of a Liquid in which a Large Number of Small Spheres are Suspended in Irregular Distribution

In the preceding discussion we have considered the case when there is suspended in a domain G, of the order of magnitude defined above, a sphere that is very small compared with this domain, and have investigated how this influenced the motion of the liquid. We will now assume that an indefinitely large number of spheres are distributed in the domain G, of similar radius and actually so small that the volume of all the spheres together is very small compared with the domain G. Let the number of spheres present in unit volume be n, where n is sensibly constant everywhere in the liquid.

We will now start once more from the motion of a homogeneous liquid, without suspended spheres, and consider again the most general motion of dilatation. If no spheres are present, by suitable choice of the co-ordinate system we can express the velocity components u_0, v_0, w_0, in the arbitrarily-chosen point x, y, z in the domain G, by the equations

$$u_0 = Ax,$$
$$v_0 = By,$$
$$w_0 = Cz,$$

where $\qquad A = B + C = 0.$

Now a sphere suspended at the point $x_\nu, y_\nu z_\nu$, will affect this motion in a manner evident from the equation (6). Since we have assumed that the average distance between neighbouring spheres is very great compared with their radius, and consequently the additional velocity-components originating from all the suspended spheres together are very small compared with u_0, v_0, w_0, we get for the velocity-components u, v, w in the liquid, taking into account the suspended spheres and neglecting terms of higher orders—

$$(8) \begin{cases} u = Ax - \Sigma\left\{\dfrac{5}{2}\dfrac{P^3}{\rho_\nu^2}\dfrac{\xi_\nu(A\xi_\nu^2 + B\eta_\nu^2 + C\zeta_\nu^2)}{\rho_\nu^3}\right. \\ \qquad \left. - \dfrac{5}{2}\dfrac{P^3}{\rho_\nu^4}\dfrac{\xi_\nu(A\xi_\nu^2 + B\eta_\nu^2 + C\zeta_\nu^2)}{\rho_\nu^3} + \dfrac{P^5}{\rho_\nu^4}\dfrac{A\xi_\nu}{\rho_\nu}\right\}, \\[2ex] v = By - \Sigma\left\{\dfrac{5}{2}\dfrac{P^3}{\rho_\nu^2}\dfrac{\eta_\nu(A\xi_\nu^2 + B\eta_\nu^2 + C\zeta_\nu^2)}{\rho_\nu^3}\right. \\ \qquad \left. - \dfrac{5}{2}\dfrac{P^5}{\rho_\nu^4}\dfrac{\eta_\nu(A\xi_\nu^2 + B\eta_\nu^2 + C\zeta_\nu^2)}{\rho_\nu^3} + \dfrac{P^5}{\rho_\nu^4}\dfrac{B\eta_\nu}{\rho_\nu}\right\}, \\[2ex] w = Cz - \Sigma\left\{\dfrac{5}{2}\dfrac{P^3}{\rho_\nu^2}\dfrac{\zeta_\nu(A\xi_\nu^2 + B\eta_\nu^2 + C\zeta_\nu^2)}{\rho_\nu^3}\right. \\ \qquad \left. - \dfrac{5}{2}\dfrac{P^5}{\rho_\nu^4}\dfrac{\zeta_\nu(A\xi_\nu^2 + B\eta_\nu^2 + C\zeta_\nu^2)}{\rho_\nu^3} + \dfrac{P^5}{\rho_\nu^4}\dfrac{C\zeta_\nu}{\rho_\nu}\right\}, \end{cases}$$

where the summation is extended over all spheres in the domain G, and we put

$$\xi_\nu = x - x_\nu,$$
$$\eta_\nu = y - y_\nu, \qquad \rho_\nu = \sqrt{\xi_\nu{}^2 + \eta_\nu{}^2 + \zeta_\nu{}^2},$$
$$\zeta_\nu = z - z_\nu.$$

x_ν, y_ν, z_ν are the co-ordinates of the centre of the sphere. Further, we conclude from the equations (7) and (7a) that the presence of each of the spheres has a result (neglecting indefinitely small quantities of a higher order) (23) in an increase of the heat production per unit volume, and that the energy per unit volume transformed into heat in the domain G has the value

$$W = 2\delta^2 k + n\delta^2 k\Phi, \qquad . \qquad . \quad (23)$$

or

(7b) $$W = 2\delta^2 k\left(\mathbf{I} + \frac{\phi}{2}\right), \qquad . \qquad . \quad (23)$$

where ϕ denotes the fraction of the volume occupied by the spheres.

From the equation (7b) the viscosity-coefficient can be calculated of the heterogeneous mixture of liquid and suspended spheres (hereafter termed briefly " mixture ") under discussion ; but we must bear in mind that A, B, C are not the values of the principal dilatations in the motion of the liquid defined by the equations (8), (23) ; we will call the principal dilatations of the mixture A^*, B^*, C^*. On the grounds of symmetry it follows that the principal directions of dilatation of the mixture are parallel to the directions of the principal dilatations A, B, C, and therefore to the co-ordinate axes. If we write the equations (8) in the form

$$u = Ax + \Sigma u_\nu,$$
$$v = By + \Sigma v_\nu,$$
$$w = Cz + \Sigma w_\nu,$$

we get

$$A^* = \left(\frac{\partial u}{\partial x}\right)_{x=0} = A + \Sigma\left(\frac{\partial u_\nu}{\partial x}\right)_{x=0} = A - \Sigma\left(\frac{\partial u_\nu}{\partial x_\nu}\right)_{x=0}.$$

If we exclude from our discussion the immediate neighbourhood of the single spheres, we can omit the second and third terms of the expressions for u, v, w, and obtain when $x = y = z = 0$:—

$$(9) \begin{cases} u_\nu = -\dfrac{5}{2}\dfrac{P^3}{r_\nu^2}\dfrac{x_\nu(Ax_\nu^2 + By_\nu^2 + Cz_\nu^2)}{r_\nu^3}, \\[2mm] v_\nu = -\dfrac{5}{2}\dfrac{P^3}{r_\nu^2}\dfrac{y_\nu(Ax_\nu^2 + By_\nu^2 + Cz_\nu^2)}{r_\nu^3}, \\[2mm] w_\nu = -\dfrac{5}{4}\dfrac{P^3}{r_\nu^2}\dfrac{z_\nu(Ax_\nu^2 + Bz_\nu^2 + Gz_\nu^2)}{r_\nu^3} \end{cases}$$

where we put

$$r_\nu = \sqrt{x_\nu^2 + y_\nu^2 + z_\nu^2} > 0.$$

We extend the summation throughout the volume of a sphere K of very large radius R, whose centre lies at the origin of the co-ordinate system. If we assume further that the irregularly distributed spheres are now evenly distributed and introduce an integral in place of the summation, we obtain

$$A^* = A - n \int_K \frac{\partial u_\nu}{\partial x_\nu} dx_\nu dy_\nu dz_\nu,$$

$$= A - n \int \frac{u_\nu x_\nu}{r_\nu} ds \qquad . \qquad . \quad (27)$$

where the last integration is to be extended over the surface of the sphere K. Having regard to (9) we find

$$A^* = A - \frac{5}{2} \frac{P^3}{R^6} n \int x_0{}^2 (A x_0{}^2 + B y_0{}^2 + C z_0{}^2) ds$$

$$= A - n \left(\frac{4}{3} P^3 \pi \right) A = A (1 - \phi).$$

By analogy

$$B^* = B(1 - \phi),$$
$$C^* = C(1 - \phi).$$

We will put

$$\delta^{*2} = A^{*2} + B^{*2} + C^{*2},$$

then neglecting indefinitely small quantities of higher order,

$$\delta^{*2} = \delta^2 (1 - 2\phi).$$

We have found for the development of heat per unit of time and volume

$$W^* = 2\delta^2 k\left(1 + \frac{\phi}{2}\right) \quad . \quad . \quad (23)$$

Let us call the viscosity-coefficient of the mixture k^*, then

$$W^* = 2\delta^{*2}k^*.$$

From the last three equations we obtain (neglecting indefinitely small quantities of higher order)

$$k^* = k(1 + 2 \cdot 5\phi) \quad . \quad . \quad (23)$$

We reach, therefore, the result :—

If very small rigid spheres are suspended in a liquid, the coefficient of internal friction is thereby increased by a fraction which is equal to $2 \cdot 5$ times the total volume of the spheres suspended in a unit volume, provided that this total volume is very small.

§ 3. On the Volume of a Dissolved Substance of Molecular Volume Large in Comparison with that of the Solvent

Consider a dilute solution of a substance which does not dissociate in the solution. Suppose that a molecule of the dissolved substance is large compared with a molecule of the solvent ; and can be thought of as a rigid sphere of radius P. We can then apply the result obtained in Paragraph 2.

If k^* be the viscosity of the solution, k that of the pure solvent, then

$$\frac{k^*}{k} = 1 + 2 \cdot 5\phi, \qquad \cdot \qquad \cdot \qquad (23)$$

where ϕ is the total volume of the molecules present in the solution per unit volume.

We will calculate ϕ for a 1 per cent. aqueous sugar solution. According to the observations of Burkhard (Landolt and Börnstein Tables) $k^*/k = 1 \cdot 0245$ (at 20° C.) for a 1 per cent. aqueous sugar solution ; therefore $\phi = 0 \cdot 0245$ for (approximately) 0·01 gm. of sugar. A gram of sugar dissolved in water has therefore the same effect on the viscosity as small suspended rigid spheres of total volume 0·98 c.c. (23)

We must recollect here that 1 gm. of solid sugar has the volume 0·61 c.c. We shall find the same value for the specific volume s of the sugar present in solution if the sugar solution is looked upon as a mixture of water and sugar in a dissolved form. The specific gravity of a 1 per cent. aqueous sugar solution (referred to water at the same temperature) at 17·5° is 1·00388. We have then (neglecting the difference in the density of water at 4° and at 17·5°)—

$$\frac{1}{1 \cdot 00388} = 0 \cdot 99 + 0 \cdot 01s.$$

Therefore $\qquad\qquad s = 0 \cdot 61.$

While, therefore, the sugar solution behaves, as to its density, like a mixture of water and solid sugar, the effect on the viscosity is one and one-half times greater than would have resulted from the suspension of an equal mass of sugar. It appears to me that this result can hardly be explained in the light of the molecular theory, in any other manner than by assuming that the sugar molecules present in solution limit the mobility of the water immediately adjacent, so that a quantity of water, whose volume is approximately one-half (23) the volume of the sugar-molecule, is bound on to the sugar-molecule.

We can say, therefore, that a dissolved sugar molecule (or the molecule together with the water held bound by it respectively) behaves in hydrodynamic relations as a sphere of volume $0.98 \cdot 342/N$ c.c. (23), where 342 is the molecular weight of sugar and N the number of actual molecules in a gram-molecule.

§ 4. On the Diffusion of an Undissociated Substance in Solution in a Liquid

Consider such a solution as was dealt with in Paragraph 3. If a force K acts on the molecule, which we will imagine as a sphere of radius P, the molecule will move with a velocity ω which is determined by P and the viscosity k of the solvent.

That is, the equation holds :—(*)

(1) $$\omega = \frac{k}{6\pi kP} \qquad . \qquad . \qquad . \quad (6)$$

We will use this relation for the calculation of the diffusion-coefficient of an undissociated solution. If p is the osmotic pressure of the dissolved substance, which is looked upon as the only force producing motion in the dilute solution under consideration, then the force exerted in the direction of the X-axis on the dissolved substance per unit volume of the solution $= - dp/dx$. If there are ρ grams in a unit volume and m is the molecular weight of the dissolved substance, N the number of actual molecules in a gram-molecule, then $(\rho/m)N$ is the number of (actual) molecules in a unit of volume, and the force acting on a molecule as a result of the fall in concentration will be

(2) $$K = - \frac{m}{\rho N} \frac{\partial p}{\partial x}.$$

If the solution is sufficiently dilute, the osmotic pressure is given by the equation

(3) $$p = \frac{R}{m} \rho T,$$

(*) G. Kirchhoff, "Lectures on Mechanics," Lect. 26 (22).

where T is the absolute temperature and $R = 8\cdot31 . 10^7$. From the equations (1), (2), and (3) we obtain for the velocity of movement of the dissolved substance

$$\omega = -\frac{RT}{6\pi k}\frac{1}{NP}\frac{\partial\rho}{\partial x}.$$

Finally, the weight of substance passing per unit of time across unit area in the direction of the X-axis will be

$$(4) \qquad \omega\rho = -\frac{RT}{6\pi k}\cdot\frac{1}{NP}\frac{\partial\rho}{\partial x}.$$

We obtain therefore for the diffusion coefficient D—

$$D = \frac{RT}{6\pi k}\cdot\frac{1}{NP}.$$

Accordingly, we can calculate from the diffusion-coefficient and the coefficient of viscosity of the solvent, the value of the product of the number N of actual molecules in a gram-molecule and of the hydrodynamically-effective radius P of the molecule.

In this calculation osmotic pressure is treated as a force acting on the individual molecules, which evidently does not correspond with the conceptions of the kinetic-molecular theory, since, according to the latter, the osmotic pressure in

the case under discussion must be thought of as
a virtual force only. However, this difficulty
vanishes if we reflect that (dynamic) equilibrium
with the (virtual) osmotic forces, which correspond
to the differences in concentration of the solution,
can be established by the aid of a numerically
equal force acting on the single molecules in the
opposite direction ; as can easily be established
following thermodynamic methods.

Equilibrium can be obtained with the osmotic

force acting on unit mass, $-\dfrac{1}{\rho}\dfrac{\partial \rho}{\partial x}$, by the force $- Px$

(applied to the individual solute molecules) if

$$-\frac{1}{\rho}\frac{\partial \rho}{\partial x} - Px = 0.$$

If we imagine, therefore, two mutually eliminating systems of forces Px and $- Px$ applied to the
dissolved substance (per unit mass), then $-Px$
establishes equilibrium with the osmotic pressure
and only the force Px, numerically equal to the
osmotic pressure, remains over as cause of motion.
Thus the difficulty mentioned is overcome.(*)

(*) A detailed statement of this train of thought will be
found in *Ann. d. Phys.*, **17**, 1905, p. 549.

§ 5. Determination of Molecular Dimensions
with the Help of the Relations already
Obtained

We found in Paragraph 3

$$\frac{k^*}{k} = 1 + 2 \cdot 5 \phi = 1 + 2 \cdot 5 n \cdot \frac{4}{3} \pi P^3 \qquad (23)$$

where n is the number of solute molecules per unit
volume and P the hydrodynamically-effective
radius of the molecule.   If we bear in mind that

$$\frac{N}{n} = \frac{\rho}{m}$$

where ρ is the mass of the dissolved substance
present in unit volume and m is its molecular
weight, we obtain

$$NP^3 = \frac{3}{10\pi} \frac{m}{\rho} \left(\frac{k^*}{k} - 1 \right).$$

On the other hand, we found in § 4

$$NP = \frac{RT}{6\pi k} \frac{1}{D}.$$

These two equations put us in the position to
calculate each of the quantities P and N, of which
N must show itself to be independent of the nature
of the solvent, of the solute and of the temperature, if our theory is to correspond with the facts.

We will carry out the calculation for an aqueous sugar solution. Firstly, it follows from the data given above for the viscosity of sugar solution at 20° C.

$$NP^3 = 80 \quad . \quad . \quad . \quad (23)$$

According to the researches of Graham (calculated out by Stephan), the diffusion-coefficient of sugar in water at 9·5° is 0·384, if the day is taken as unit of time. The viscosity of water at 9·5° is 0·0135. We will insert these data in our formula for the diffusion-coefficient, although they were obtained with 10 per cent. solutions, and it is not to be expected that our formula will be precisely valid at so high a concentration. We get

$$NP = 2·08 . 10^{16}.$$

It follows from the values found for NP^3 and NP, if we ignore the difference in P at 9·5° and 20°, that

$$P = 6·2 . 10^{-8} \text{ cm.} \quad . \quad . \quad (23)$$
$$N = 3·3 . 10^{23}.$$

The value found for N agrees satisfactorily, in order of magnitude, with the values obtained by other methods for this quantity.

Berne, 30 *April*, 1905.

(Received, 19 *August*, 1905.)

Supplement

In the new edition of Landolt and Börnstein's " Physical-Chemical Tables " will be found very useful data for the calculation of the size of the sugar molecule, and the number N of the actual molecules in a gram-molecule. Thovert found (Table, p. 372) for the diffusion-coefficient of sugar in water at 18·5° C. and the concentration 0·005 mol./litre the value 0·33 cm.²/day.   From a table (p. 81), with the results of observations made by Hosking, we find by interpolation that in dilute sugar solutions an increase in the sugar-content of 1 per cent. at 18·5° C. corresponds to an increase of the viscosity of 0·00025.   Utilizing these data, we find

$$P = 0.49 \cdot 10^{-6} \text{ mm.}$$

and

$$N = 6.56 \cdot 10^{23}. \qquad (23), (28)$$

Berne, *January*, 1906.

IV

THEORETICAL OBSERVATIONS ON THE BROWNIAN MOTION

(From *Zeit. f. Elektrochemie*, **13**, 1907, pp. 41-42)

IN connection with the researches of Svedberg,(29) recently published in the *Zeit. f. Elektrochemie*, on the motion of small suspended particles, it appears to me desirable to point out some properties of this motion indicated by the molecular theory of heat. I hope I may be able by the following to facilitate for physicists who handle the subject experimentally the interpretation of their observations as well as the comparison of the latter with the theory.

1. From the molecular theory of heat we can calculate the mean value of the instantaneous velocity which a particle may have at the absolute temperature T. Thus the kinetic energy of the motion of the centre of gravity of a particle is independent of the size and nature of the particle and independent of the nature of its environment,

e.g. of the liquid in which the particle is suspended :
this kinetic energy is equal to that of a monatomic
gas molecule.   The mean velocity $\sqrt{\overline{v^2}}$ of the particle of mass m is therefore determined by the
equation

$$m\frac{\overline{v^2}}{2} = \frac{3}{2}\frac{RT}{N},$$

where $R = 8 \cdot 3 \cdot 10^7$, T is the absolute temperature, and N the number of the actual molecules
in a gram-molecule (approximately $6 \cdot 10^{23}$, (19)).
We will calculate the value of $\sqrt{\overline{v^2}}$, as well as
other quantities discussed in the following, for
particles in colloidal platinum solutions such as
Svedberg has investigated.   For these particles
we have to put $m = 2 \cdot 5 \cdot 10^{-15}$, so that we get
for $T = 292$

$$\sqrt{\overline{v^2}} = \sqrt{\frac{3RT}{mN}} = 8 \cdot 6 \text{ cm./sec.}$$

2. We will now examine whether there is any
prospect of actually observing this enormous velocity of a suspended particle.

If we knew nothing of the molecular theory of
heat, we should expect the following to happen.
Suppose that we impart to a particle suspended
in a liquid a certain velocity by an impulsive force
applied to it from without ; then this velocity

will die away rapidly on account of the friction
of the liquid. We will ignore the inertia of the
latter and note that the resistance that the particle moving with the velocity v experiences is
$6\pi kPv$,(6) where k is the viscosity of the liquid
and P the radius of the particle. We obtain the
equation

$$m\frac{dv}{dt} = -\ 6\pi kPv.$$

From this it follows that for the time θ in which
the velocity falls to a tenth of its original value—

$$\theta = \frac{m}{0\cdot 434\ .\ 6\pi kP}.\qquad .\qquad . \quad (30)$$

For the platinum particles (in water), mentioned
before, we have to put $P = 2\cdot5\ .\ 10^{-6}$ cm. and
$k = 0\cdot01$, so that we get (*)

$$\theta = 3\cdot3\ .\ 10^{-7} \text{ seconds.}$$

If we turn back again to the molecular theory
of heat, we have to modify this conception. In
fact, we must now also assume that the particle
nearly completely loses its original velocity in the
very short time θ through friction. But, at the
same time, we must assume that the particle gets

(*) For particles of " microscopic " size θ is appreciably
greater, since θ is proportional to the square of the radius
of the particles, other conditions being the same.

new impulses to movement during this time by some process that is the inverse of viscosity, so that it retains a velocity which on an average is equal to $\sqrt{\overline{v^2}}$. But since we must imagine that direction and magnitude of these impulses are (approximately) independent of the original direction of motion and velocity of the particle, we must conclude that the velocity and direction of motion of the particle will be already very greatly altered in the extraordinary short time θ, and, indeed, in a totally irregular manner.

It is therefore impossible—at least for ultra-microscopic particles—to ascertain $\sqrt{\overline{v^2}}$ by observation.

3. If we confine ourselves to the investigation of the lengths of path, or, more precisely expressed, the changes in position in times τ, which are substantially greater than θ, then from the molecular theory of heat

$$\sqrt{\overline{\lambda x^2}} = \sqrt{\tau}\sqrt{\frac{RT}{N}\frac{1}{3\pi kP}}, \qquad . \qquad . \quad (31)$$

if the change in the X-co-ordinate of the particle that has taken place in the time τ is indicated by λx. (42)

For the mean velocity in the interval of time τ we can define the quantity

$$\frac{\sqrt{\overline{\lambda x^2}}}{\tau} = \frac{w}{\sqrt{\tau}}$$

where for brevity we put

$$\sqrt{\frac{RT}{N} \frac{1}{3\pi kP}} = w.$$

But this mean velocity is the greater, the smaller τ is ; so long as τ is great compared with θ, the velocity does not approach any limiting value as τ decreases.

Since an observer operating with definite means of observation in a definite manner can never perceive the actual path traversed in an arbitrarily small time, a certain mean velocity will always appear to him as an instantaneous velocity. But it is clear that the velocity ascertained thus corresponds to no objective property of the motion under investigation—at least, if the theory corresponds to the facts.

Berne, *January*, 1907.

(Received, 22 *January*.)

THE ELEMENTARY THEORY OF THE BROWNIAN (*) MOTION

(From the *Zeit. für Elektrochemie*, **14**, 1908, pp. 235-239)

PROF. R. LORENTZ has called to my attention, in a verbal communication, that an elementary theory of the Brownian motion would be welcomed by a number of chemists. Acting on this invitation, I present in the following a simple theory of this phenomenon. The train of thought conveyed is briefly as follows.

First we investigate how the process of diffusion in an undissociated dilute solution depends on the distribution of osmotic pressure in the solution and on the mobility of the dissolved substance in the solvent. We thus obtain an expression for the diffusion-coefficient in the case when a mole-

(*) We mean by Brownian motion that irregular movement which small particles of microscopic size carry out when suspended in a liquid. Refer e.g. to Th. Svedberg, *Zeit. f. Flektrochem.*, **12**, 47 and 51, 1906.

cule of the dissolved substance is great compared with a molecule of the solvent, in which expression no quantities dependent on the nature of the solution appear except the viscosity of the solvent and the diameter of the solute molecules.

After this we relate the process of diffusion to the irregular motions of the solute molecules; and find how the average magnitude of these irregular motions of the solute molecules can be calculated from the diffusion-coefficient, and therefore, with the help of the results indicated above, from the viscosity of the solvent and the size of the solute molecules. The result so obtained holds not only for actual dissolved molecules, but also for any small particles suspended in the liquid.

§ 1. DIFFUSION AND OSMOTIC PRESSURE

Suppose the cylindrical vessel Z (Fig. 93) filled with a dilute solution. The interior of Z is divided by a movable piston k, which forms a semipermeable partition, into two parts A and B. If the concentration of the solution in A is greater than that in B, an exterior force, directed towards the left, must be applied to the piston in order to retain it in equilibrium; this force is in fact equal to the difference of the two osmotic pressures

which the dissolved substance exerts on the piston
on the left and on the right side respectively. If
this external force is not allowed to act on the
piston, it will move under the influence of the
greater osmotic pressure of the solution present in
A so far to the right that the concentrations in
A and *B* no longer differ. From this consideration it follows that it is the forces of osmotic
pressure that bring about the equalization of the

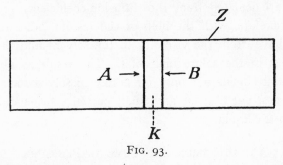

FIG. 93.

concentrations in diffusion ; for we can prevent
diffusion, that is, an equalization of concentration, by balancing the osmotic differences, which
correspond to the differences of concentration,
by external forces acting on semi-permeable par-
titions. It has long been realized that the osmotic pressure can be looked upon as the driving
force in diffusion phenomena. It is familiar that
Nernst (32) made this the foundation of his investi-

gations into the connection between ionic mobility, diffusion-coefficient, and E.M.F. in concentration cells.

Suppose a diffusion process is taking place within the cylinder Z (Fig. 94), of unit area of cross-section, in the direction of the axis of the cylinder. We will investigate first the osmotic forces given rise to by the motion—due to dif-

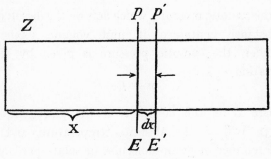

FIG. 94.

fusion—of the dissolved substance contained between the planes E and E' at an indefinitely short distance from one another. The osmotic pressure force p acts on the surface E of the layer from left to right, the force p' acts on the surface E' from right to left ; the resultant of the pressure forces is therefore

$$p - p'.$$

We will now call x the distance of the surface E from the left end of the vessel, $x + dx$ the distance of the surface E' from that end ; then dx is the volume of the layer of liquid in question. Since $p - p'$ is the osmotic pressure which acts on the volume dx of the dissolved substance, then—

$$K = \frac{p - p'}{dx} = -\frac{p' - p}{dx} = -\frac{dp}{dx}$$

is the osmotic pressure, which acts on the dissolved substance contained in unit volume. Since, further, the osmotic pressure is given by the equation

$$p = R\nu \qquad . \qquad . \qquad . \qquad (2)$$

where R is the constant of the gas-equation ($8 \cdot 31 \cdot 10^7$), T the absolute temperature, and ν the number of gram-molecules of solute per unit volume, we get, finally, for the osmotic force K acting on the dissolved substance per unit volume the expression

$$(1) \qquad\qquad K = -RT \frac{d\nu}{dx}.$$

Now, in order to be able to calculate the motions, due to diffusion, to which these active forces can give rise, we must know how great a resistance the solvent offers to a movement of the dissolved substance. If an active force K acts on a molecule, this will impart to the molecule a proportional velocity v, according to the equation

$$(2) \qquad\qquad v = \frac{k}{\Re},$$

where $\Re$ is a constant, which we will call the frictional resistance of the molecule. This frictional resistance cannot in general be deduced theoretically. But when the dissolved molecule can be looked upon approximately as a sphere, which is large compared with a molecule of the solvent, we may ascertain the frictional resistance of the solute molecule according to the methods of ordinary hydrodynamics, which do not take account of the molecular constitution of the liquid. Within the limits of valid application of ordinary hydrodynamics, for a sphere moving in a liquid the equation (2) holds, where we put

$$(3) \qquad\qquad \Re = 6\pi\eta\rho. \qquad . \qquad . \qquad (6)$$

Here η denotes the coefficient of viscosity of the liquid, ρ the radius of the sphere. If it can be assumed that the molecules of a solute are approximately spherical and are large compared with the molecules of the solvent, the equation (3) may be applied to the single solute molecules.

We can now estimate the mass of a solute

diffusing across a cross-section of the cylinder per
unit of time.   There are ν gram-molecules present
in the unit volume, therefore νN actual molecules, where N signifies the number of actual
molecules in a gram-molecule.   If a force K is
distributed over these νN molecules contained in
the unit volume, it will impart to these a νN-times
smaller velocity than it is able to impart to a
single molecule, if acting on the latter alone.
Reverting to equation (2) : for the velocity v,
which the force K is able to impart to the νN
molecules, we obtain the expression

$$v = \frac{1}{\nu N} \cdot \frac{K}{\mathfrak{R}}.$$

In the case under consideration, K is equal to
the osmotic force previously calculated, which acts
on the νN molecules in a unit volume ; so that we
obtain from the above, using equation (1),

$$(4) \qquad v\nu = -\frac{RT}{N} \cdot \frac{1}{\mathfrak{R}} \cdot \frac{dv}{dx}.$$

On the left-hand side we have the product of
the concentration ν of the solute, and of the
velocity, with which the latter substance will be
moved forward by the process.   This product
therefore represents the mass of the dissolved
substance (in gram-molecules) which is carried

per second by diffusion through unit area of cross-section. The multiplier of dv/dx on the right-hand side of this equation is therefore (*) nothing else but the coefficient of diffusion D of the solution in question. We have, therefore, in general

$$(5) \qquad D = \frac{RT}{N} \cdot \frac{1}{\mathfrak{R}},$$

and, in the case when the diffusing molecules can be looked upon as spherical, and large compared to the molecules of the solvent, introducing equation (3),

$$(5a) \qquad D = \frac{RT}{N} \frac{1}{6\pi\eta\rho} \qquad \cdot \qquad \cdot \qquad (33)$$

In the last case, therefore, the coefficient of diffusion depends upon no other constants characteristic of the substance in question but the viscosity η of the solvent and the radius ρ of the molecule. (†)

(*) It is to be noted that the numerical value of the coefficient of diffusion is independent of the unit taken for concentration.

(†) This equation enables the radius of large molecules to be deduced approximately from the coefficient of diffusion, when the latter is known ; it is then

$$\rho = \frac{RT}{6\pi N\eta} \cdot \frac{1}{D}$$

where $R = 8.31 \cdot 10^7$, $N = 6 \cdot 10^{23}$. Of course, a degree of uncertainty of some 50 per cent. is involved in the

§ 2. DIFFUSION AND IRREGULAR MOTION OF THE MOLECULES

The molecular theory of heat affords a second point of view, from which the process of diffusion can be considered. The process of irregular motion which we have to conceive of as the heat-content of a substance will operate in such a manner that the single molecules of a liquid will alter their positions in the most irregular manner thinkable. This wandering about of the molecules of the solute—fortuitous to a certain extent—in a solution will have as a result that an originally non-uniform distribution of concentration of the solute will gradually give place to a uniform one.

We will now examine this process somewhat more narrowly, whilst we confine ourselves again to the case considered in § 1, fixing our attention on the diffusion in one direction only, namely, in the direction of the axis (x-axis) of the cylinder Z.

We will imagine that we know the x-co-ordinates of all solute molecules at a certain time t, and also at the time $t + \tau$, where τ indicates an interval of time so short that the relation of the concentrations of our solution alters only very slightly during this interval. During this time τ the

value of N. (34)   This relation should be of importance for the determination of the approximate dimensions of the molecules in colloidal solutions.

x-co-ordinates of the first solute molecule will have changed, through the irregular thermal motion, by a certain amount Δ_1, that of the second molecule by Δ_2, etc. These displacements, Δ_1, Δ_2, etc., will be partly negative (towards the left), partly positive (towards the right). The magnitude of this displacement will, further, be

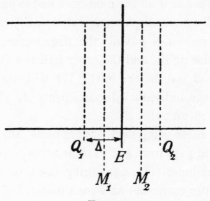

FIG. 95.

different for the individual molecules. But since, as before, we presuppose a dilute solution, this displacement is controlled only by the surrounding solvent, and not to a sensible extent by the rest of the solute molecules ; hence, in portions of the solution of different concentrations these displacements Δ will be on an average of equal magnitude, just as frequently positive as negative.

We will now see how large the mass of the substance turns out to be, which diffuses in the time τ through unit area of cross-section of a solution, when the magnitude is known of the displacement $\varDelta$ in the direction of the axis of the cylinder, which the solute molecules experience on an average. To simplify this investigation, we will make our calculations as if all the molecules had experienced an equally great displacement $\varDelta$, actually one-half of the molecules having the displacement $+ \varDelta$ (i.e. to the right), and the other half the displacement $- \varDelta$ (i.e. to the left). We will, therefore, replace the individual displacements $\varDelta_1$, $\varDelta_2$, etc., by their mean value $\varDelta$.

With these simplified assumptions, there will be able to pass from left to right across a plane E of our cylinder (Fig. 95) during the time τ, only such solute molecules as were situated before the period of the time τ on the left of E. and at a distance from E which is less than $\varDelta$. These molecules are all situated between the planes Q_1 and E (Fig. 95). But since only half of these molecules experience the displacement $+ \varDelta$, only half of them will also pass across the plane E. The half of the solute substance situated between Q and E is, however, when expressed in gram-molecules, equal to

$$\frac{1}{2}\nu_1\varDelta,$$

where ν_1 is the mean concentration in the volume Q_1E, i.e. the concentration in the middle layer M_1. Since the cross-section is unity, $\varDelta$ is the volume included between Q_1 and E, which, when multiplied by the mean concentration, gives the amount of the solute in gram-molecules contained in this volume.

By similar reasoning, it follows that the mass of the solute which passes across E from right to left in the time τ is equal to

$$\frac{1}{2}\nu_2\varDelta$$

where ν_2 denotes the concentration in the middle layer M_2. The quantity of substance which diffuses across from left to right during the time τ is then obviously equal to the difference of these two expressions, therefore equal to

(6) $$\frac{1}{2}\varDelta(\nu_1 - \nu_2).$$

ν_1 and ν_2 are the concentrations in two cross-sections which are separated by the very small distance $\varDelta$. Again, if we denote by x the distance of a cross-section from the left cylinder-end, according to the definition of a differential,

$$\frac{v_2 - v_1}{\varDelta} = \frac{dv}{dx},$$

thence

$$v_1 - v_2 = -\varDelta\frac{dv}{dx},$$

so that the quantity of the substance which diffuses across E during time τ is also equal to

(6a)
$$-\frac{1}{2}\varDelta^2\frac{dv}{dx}.$$

The quantity of the substance (expressed in gram-molecules) which diffuses across E in a unit of time is therefore—

$$-\frac{1}{2}\frac{\varDelta^2}{\tau}\frac{dv}{dx}.$$

We have thereby obtained a second value for the coefficient of diffusion D. It is

(7)
$$D = \frac{1}{2}\frac{\varDelta^2}{\tau},$$

where $\varDelta$ signifies the length of path described on an average (*) by a solute molecule during the time τ in the direction of the x-axis.

(*) More accurately, $\varDelta$ should be put equal to the square root of the mean of the squares of the individual displacements $\varDelta_1{}^2$, $\varDelta_2{}^2$, etc. We should therefore write, with greater accuracy, $\sqrt{\varDelta^2}$ in place of $\varDelta$.

Solving the equation (7) for $\varDelta$, we obtain

(7a) $$D = \sqrt{2D}\sqrt{\tau}.$$

§ 3. Movement of the Single Molecules : Brownian Motion

If in the equations (5) and (7) we put the values given for the coefficients of diffusion equal to one another, we obtain, solving for $\varDelta$,

(8) $$\varDelta = \sqrt{\frac{2RT}{N\Re}}\sqrt{\tau}.$$

We see from this formula that the path described by a molecule on an average is not proportional to the time,(*) but proportional to the square root of the time.   This follows from the fact that the paths described during two consecutive unit time-intervals are not always to be added, but just as frequently have to be subtracted.   We can calculate the displacement of the molecule resulting on an average from the irregular molecular motion : by means of equation (7a) from the coefficient of diffusion, by means of equation (8) from the resistance which is offered to a forced motion of velocity $v = 1$.

(*) Compare A. Einstein, *Z. f. Elektroch.*, **13** (1907) ; and pp. 63-67 of this volume.

In the case when the solute molecule is large compared to the molecule of the solvent, and is spherical, we can put the value of R given in equation (3) in equation (8), so that we obtain

(8a) $$\varDelta = \sqrt{\frac{RT}{N}\frac{1}{3\pi\eta\rho}} \cdot \sqrt{\tau} \qquad . \qquad . \quad (31)$$

This equation enables the mean displacement $\varDelta$ (*) to be calculated from the temperature T, the viscosity of the solute η and the radius ρ of the molecule.

According to the molecular kinetic conception, there exists no essential difference between a solute molecule and a suspended particle. We will therefore consider equation (8a) as also valid for the case where we deal with any kind of small suspended spherical particles.

We will calculate the length of path $\varDelta$ which a particle of 1μ diameter describes on an average in one second in a certain direction in water at room temperature. We have to put

$R = 8\cdot31 \,.\, 10^7.$ $\eta = 0\cdot0135.$

$T = 290.$ $\rho = 0\cdot5 \,.\, 10^{-4}.$

$N = 6\cdot10^{23}.$ $\tau = 1.$

(*) More accurately the square root of the mean value of $\varDelta^2$.

We obtain

$$\Delta = 0 \cdot 8 \times 10^{-4} \text{ cm.} = 0 \cdot 8\mu.$$

This number is subject to an error of some $\pm$ 25 per cent. on account of the limited degree of accuracy with which N is known. (34)

It is of interest to compare the mean individual motions of microscopic particles calculated in this manner, with those of solute molecules and of ions respectively. For an undissociated dissolved substance, whose coefficient of diffusion is known, we can calculate Δ from the equation (7a). For sugar at room temperature

$$D = \frac{0 \cdot 33}{24 \cdot 60 \cdot 60}.$$

Hence we calculate from equation (7a) for $\tau = 1$

$$\Delta = 27 \cdot 6\mu.$$

One can deduce from the number N and the molecular volume of solid sugar that the diameter of a molecule of sugar is of the order of magnitude of $1\mu\mu$, therefore about a thousand times smaller than the diameter of the particle considered above. From the equation (8a) we must therefore expect that Δ for sugar will be about $\sqrt{1000}$ times greater than for the particles of 1μ diameter. This is actually approximately correct, as can be seen.

From the equation (8) we can calculate the value of l for ions from their velocity of migration. l is equal to the quantity of electricity in coulombs, which passes across a square centimeter in one second for a concentration $\nu = 1$ of the ion in question, and for a potential gradient of 1 volt per centimeter. In the case we are considering, the velocity v of the ionic motion (in cm./sec.) is evidently determined by the equation

$$l = v \cdot 96000.$$

Since, further, 1 volt is equivalent to 10^8 electromagnetic units, and the charge of a (univalent) ion is equal to $9600/N$ electromagnetic units, the force k acting on one ion in the case considered is

$$k = \frac{10^8 \cdot 9600}{N}.$$

If we put in equation (2) this value of k, and the value of v obtained in the former equation,

$$v = \frac{l}{96000},$$

we get

$$\Re = \frac{k}{v} = \frac{10^8 \cdot 9600 \cdot 96000}{lN}.$$

This expression also holds, with the usual definition of l, for polyvalent ions. Introducing this value for R in equation (8), we get

$$\Delta = 4 \cdot 25 \cdot 10^{-5} \sqrt{lT\tau}.$$

The formula gives for room temperature, and $\tau = 1$:—

Ion	l.	Δ.
H	300	125μ
K	65	58μ
Diisoamyl-ammonium, $C_{10}H_{24}N$.	24	35μ

NOTES

(1) p. 1.—The so-called "Brownian move-ment" was described for the first time in the year 1828 by the botanist Robert Brown.[1] In investigating the pollen of different plants he observed that this became dispersed in water in a great number of small particles, which were perceived to be in uninterrupted and irregular "swarming" motion. As the phenomenon re-peated itself with all possible kinds of organic substances, he believed that he had found in these particles the "primitive molecule" of living matter. He found later that the particles of every kind of inorganic substance presented the same phenomenon, so that he drew the conclusion that all matter was built up of "primitive mole-cules."

Of the authors who carried out investigations on the Brownian movement before Einstein, we will mention the following : Regnault (1858) thought that the motion was caused by irregular heating

[1] *Phil. Mag.* (4), 1828, p. 161 ; *Ann. d. Phys. u. Chem.*, **14**, 294 (1828).

by incident light. Chr. Weiner (1863) concluded
that it could not have been brought about by
forces exerted by the particles on one another, nor
by temperature differences, nor by evaporation.
Cantoni and Oehl (1865) found that the movement
persisted unchanged for a whole year when the
liquid was sealed up between two cover-glasses.
S. Exner (1867) found that the movement is most
rapid with the smallest particles, and is increased
by light and heat rays. The idea of Jevons (1870)
that the phenomenon is caused by electrical forces
was denied by Dancer (1870), who showed that
electrical forces had no influence on the motion.
In 1877 Delsaux expressed for the first time the
now generally-accepted idea that the Brownian
movement has its origin in the impacts of the
molecules of the liquid on the particles. This
point of view was also expressed by Carbonelle.

The first precise investigations we owe to Gouy,[1]
who found that the motion is the more lively the
smaller the viscosity of the liquid is (as follows
also from the theory of Einstein) ; that very
considerable changes of the intensity of illumination had no influence, nor had an extraordinarily
strong electromagnetic field. He also ascribed
the motion to the effect of the thermal molecular
motions of the liquid, and found by measurement

[1] M. Gouy, *Journ. de Phys.* (2), **7**, 561, 1888.

the velocity of different particles to be about a hundred-millionth of the molecular velocity.

Ramsay in 1892 disputed the possibility of an electrical origin of the Brownian movement, and affirmed that it must give rise to a pressure, by which certain departures from the established laws of osmotic pressure could be explained. Mäede Bache, in 1894, also accepted Gouy's point of view ; while Quincke, in 1898, looked upon the motion as a result of temperature differences in the liquid.

Besides Gouy's work there is only one other investigation of a precise nature before Einstein's treatment of the problem : that carried out by F. M. Exner,[1] who challenged Quincke's assertion, and established that the velocity of the movement decreases with increase of size of the particles and increases with rise of temperature. He expressed also the view that the kinetic energy of the particle must be equal to that of a gas molecule. Since, however, he calculated the former from the observed " velocity " of the particle, which is actually much smaller than the true velocity, his results did not agree. It first became possible to verify this relation by means of measurements of the Brownian motion made according to Einstein's method.

[1] F. M. Exner, *Ann. d. Phys.*, **2**, 843, 1900.

(2) p. 2.—Van't Hoff's law.

(3) p. 5.—The formula for the entropy S of a system depending (in the manner implied in statistical mechanics) upon the variables of condition $p_1, p_2 \ldots p_n$, used in the following treatment, is derived by Einstein in the paper quoted,[1] on the foundations of statistical mechanics. The underlying idea is roughly as follows :—

It is first shown that for the case when the system under consideration stands in statistical equilibrium with a second of the same temperature, but of indefinitely large energy content, the probability for a condition of the system in question in which the parameters lie between the values $p_1 \ldots p_1 + dp_1$, $p_2 \ldots p_2 + dp_2$, $\ldots$, $p_n \ldots p_n + dp_n$, will be given by the expression

$$dW = \text{const. } e^{-2hE} dp_1 \ldots dp_n.$$

Here E denotes the energy of the system corresponding to the statistical states $p_1 \ldots p_n$, and $2h = N/RT$. This expression corresponds to the " canonic " distribution in Gibbs' statistical mechanics. The equation can also be written

(1) $$hW = e^{c - 2hE} dp_1 \ldots dp_n,$$

where the constant c is determined by the condition

(2) $$\int e^{c - 2hE} dp_1 - dp_n = 1.$$

[1] A. Einstein, *Ann. d. Phys.*, **11,** 170, 1903.

Now let the system be dependent on a definite parameter λ which we can control arbitrarily from without, as well as on the parameter p. If we carry out an indefinitely small alteration in our system by varying this parameter λ : whilst before this change equation (2) holds, after the change

$$(3) \quad \int e^{(c + dc) - 2(h + dh)(E + \Sigma\frac{\delta E}{\delta\lambda}d\lambda)} \, dp_1 \ldots dp_n = \mathrm{I} \; ;$$

from (2) and (3) it follows

$$\int (dc - 2Edh - 2h\Sigma\frac{\partial E}{\partial\lambda}d\lambda)e^{c - 2hE}dp_1 \ldots dp_n = 0.$$

Since in the process the energy E undergoes only an indefinitely small change, it follows that

$$(4) \qquad dc - 2Edh - 2h\Sigma\frac{\partial E}{\partial\lambda}d\lambda = 0.$$

But since

$$(5) \qquad dE = \Sigma\frac{\partial E}{\partial\lambda}\,d\lambda + dQ,$$

as is easily seen, where dQ indicates the quantity of heat absorbed during the process, there follows from (4) and (5)

$$2h \, . \, dQ = d(2hE - c) \; ;$$

or multiplying by $2k$

$$\frac{dQ}{T} = d\left(\frac{E}{T} - 2xc\right) = dS.$$

The expression introduced in the text for the entropy follows immediately from this in conjunction with (2).

The definition of entropy given here is in substantial agreement with the entropy formula of Boltzmann, when taken in connection with the expression (1), which has also been already given by Boltzmann as a generalization of Maxwell's law of distribution of velocities.[1]

(4) p. 8.—From expression (1) it follows that

$$dB = dW \cdot e^{-c}$$

and since from (2)

$$\int dB = e^{-c} = B,$$

it follows that

$$dW = \frac{dB}{B}.$$

(5) p. 10.—From the relation $B = J \cdot V^{*n}$ there follows

$$S = \frac{\bar{E}}{T} + \frac{R}{N} \log B = \frac{\bar{E}}{T} + \frac{R}{N} (\log J + n \log V^*),$$

thence, since J is independent of the x-co-ordinates

$$\delta S = \frac{R}{N} \frac{n}{V^*} \delta v^* = \int_0^l \frac{R}{N} \nu \delta dx = \int_0^l \frac{R}{N} \nu d\delta x = \int_0^l \frac{R}{N} \nu \frac{\partial \delta x}{\partial x} dx.$$

[1] *Vide*, amongst others, M. v. Smoluchowski : the limits of validity of the second law of thermodynamics. Lectures on " The Kinetic Theory of Matter and Electricity," Leipzig and Berlin, 1914.

We obtain by partial integration from the last part of this expression, since the variation δ vanishes at the boundaries of the domain,

$$\delta S = - \int_0^l \frac{R}{N} \frac{\partial \nu}{\partial x} \delta x dx.$$

The expression $K\nu - \frac{\partial p}{\partial \lambda}$ can also be deduced directly without this calculation, from the existence of a force of osmotic pressure which was established at the end of § 2, equilibrium with which must be maintained by the force K.

(6) p. 11.—This expression for the resistance experienced by a sphere in a uniform movement of translation through a viscous liquid was first deduced by Stokes hydrodynamically, with the assumption that the liquid adheres completely to the surface of the sphere and its velocity becomes vanishingly small : so that the velocity of motion does not exceed a certain value.

There is no doubt that when the above conditions are fulfilled Stokes' formula really gives the motion of the sphere accurately, but it is a question whether the conditions are really fulfilled in the case of the Brownian motion of very small spherical particles.

Then, on account of its derivation, the formula is only valid for the case when the hydrodynamic equations still hold, which from the Atomic point of view can only approximately be the case so long as the radii of the spheres are large compared to the free paths of the liquid molecules. This condition is actually fulfilled by the particles of visible size in liquids, but not in gases, so that it is necessary in the latter case to apply certain corrections to Stokes' formula, which can be derived by consideration of the kinetics of gases. The first correction of this type was given by E. Cunningham [1] for the case when P is comparable with the free path of the gas molecule. It appears that the expression for the velocity of the particle must be multiplied by the factor

$$1 + A \cdot l/P,$$

where A is a constant which can have values between 0·815 and 1·63, according as to whether all impacts of the molecules against the particles are elastic or inelastic in nature. According to F. Zerner, however, these limits must be corrected to 1·40 and 1·575. [2]

Experimental tests of the law of resistance in gases have been carried out by different investigators : by M. Knudsen and S. Weber [3] by variation of the gas-pressure whilst employing a fixed size of sphere, and by L. W. McKeehan [4] also for

[1] E. Cunningham, *Proc. Roy. Soc.* (A), **83**, 357.
[2] F. Zerner, *Phys. Zeit.*, **20**, 546, 1919.
[3] Knudsen and Weber, *Ann. d. Phys.*, **36**, 981, 1911.
[4] McKeehan, *Phys. Zeit.*, **12**, 707, 1911.

different sized spheres. They obtained the following empirical formula for the correction-factor A as a function of the quantity P/l:—

$$A = 0 \cdot 68 + 0 \cdot 35 e^{-1 \cdot 85 \frac{P}{l}}.$$

This formula also holds for the case when the radius of the sphere is small compared with the free path. The researches of E. Meyer and W. Gerlach,[1] and of J. Parankiewicz [2] are in agreement with this, whilst J. Roux [3] obtained values between $1 \cdot 23$ and $1 \cdot 64$. That there can be no agreement with Cunningham's law is also evident from the researches of R. Fürth [4] on the determination of mobility from the Brownian movement.

It appears, further, that it might well be assumed that the velocities of the particles involved in the Brownian movement remain below the limits for which the Stokes formula is valid. It can be taken, from an investigation of H. D. Arnold,[5] that the Stokes formula holds below a

[1] E. Meyer and W. Gerlach, *Elster-Geitel Festschrift*, Vieweg, pp. 196, etc.

[2] J. Parankiewicz, *Phys. Zeit.*, **19**, 280, 1918.

[3] J. Roux, *Ann. de Chim. et Phys.*, viii., **29**, 69, 1913.

[4] R. Fürth, *Ann. d. Phys.*, **60**, 77, 1919 ; **63**, 521, 1920.

[5] Arnold, *Phil. Mag.*, **22**, 755, 1911.

velocity V, which satisfies the inequality condition

$$\frac{P\sigma V}{k} < 0\cdot 2,$$

where σ indicates the specific gravity of the liquid. As is shown by an approximate calculation, it is, in general, scarcely to be expected that the velocities resulting from the Brownian motion could reach these upper limits.

Finally, it must be borne in mind that the Stokes formula is deduced for constant motions of translation and established experimentally for these conditions ; so that it is not impossible that considerable divergencies may occur with accelerated motions. For certain special cases of accelerated motion, the form of the law of resistance has also been determined theoretically, e.g. for the case of small pendulum vibrations ; and can also be established experimentally in a satisfactory manner. It is then a question whether it is permissible to apply the Stokes law to the Brownian motion, which in reality exhibits no regular translation, but has an irregular character. Again, if the formula cannot be applied with certainty to the single zig-zags of the Brownian motion, it can be still assumed that, on account of the irregular character of the motion, the departures from the Stokes law cancel out on an average. Einstein's deduction given here corresponds to this thought : in which there is assumed to be a statistical equilibrium between the process of diffusion and a fictitious constant force K.

In addition, the Stokes law has been proved for accelerated motions of a regular character at high frequencies in gases, and is known to hold with very close approximation up to periods of about 60 per second.[1]

A detailed discussion of all the problems suggested here is to be found in a paper by J. Weysenhoff.[2]

(7) p. 12.—It is notable that the result for D no longer contains the applied force K. This must, however, be the case, if the method introduced here is successful in its aim, since K is a completely fictitious force which has nothing to do with the process of diffusion itself. This circumstance indicates that it must also be possible to obtain the result without the introduction of the fictitious force. Such deductions have actually been carried out ; amongst others might be particularly mentioned, on account of its special simplicity, the deduction by Ph. Frank[3] with

[1] N. A. Shewhart, *Phys. Rev.* (2), **9**, 425, 1917 ; R. B. Abott, *Phys. Rev.*, **12**, 381, 1918 ; A. Snethlage, *Versl. K. Akad. v. Wetensch. Amst.*, **25**, 1173, 1917 ; R. Fürth, *Ann. d. Phys.*, **63**, 521, 1920.

[2] J. Weysenhoff, *Ann. d. Phys.*, **62**, 1, 1920.

[3] Ph. Frank, *Wiener Ber.*, **124** (2a), 1173, 1915 ; *Ann. d. Phys.*, **52**, 323, 1917.

the help of the conception of the Virial introduced by Clausius.

(8) p. 13.—The introduction of this time-interval τ forms a weak point in Einstein's argument, since it is not previously established that such a time-interval can be assumed at all. For it might well be the case that, in the observed interval of time, there was a definite dependence of the motion of the particle on the initial state.

A deduction of the formula for the Brownian motion, which does not involve this presupposition, has been given by L. S. Ornstein [1] according to a method suggested by Frau de Haas-Lorentz,[2] as well as by R. Fürth [3] (in agreement with the former) by another method.

In contrast with the formula of Einstein, given on page 17 of the text, this formula runs

$$\overline{x^2} = 2D\left(t - mB + e^{-\frac{t}{mB}}\right)$$

where $B = 1/6\pi kP$ indicates the " Mobility " of the particle and m its mass. For a sufficiently large time-interval, the formula actually comes into line with Einstein's, whilst for very short

[1] L. S. Ornstein, *Proc. Amst.*, **21**, 96, 1918.

[2] L. de Haas-Lorentz, " The Brownian Movement and some Related Phenomena," *Die Wissenschaft*, B. 52, Vieweg, 1913.

[3] R. Fürth, *Zeit. f. Phys.*, **2**, 244, 1920.

times it indicates a rectilinear and uniform motion.

As an approximate calculation shows, the Einstein formula holds for particles of a size that can be observed, under all circumstances.

(9) p. 16.—There is sought here a so-called " Source-Integral " (" Quellenintegral ") of a differential equation of the second order, that is, a solution for which the boundary condition is assumed as a definite value for the integral of the diffusion-stream over the source. The corresponding diffusion problem is : if at time $t = 0$ the concentration of the diffusing substance is everywhere zero with the exception of an indefinitely narrow space around the plane $x = 0$, but such that the whole mass of the substance is given at all times by

$$\int_{-\infty}^{+\infty} f(x, t)dx = n,$$

then the formula given for $f(x, t)$ will be the expression for the distribution of the concentration of the substance at some later time t and at any point x.[1]

Similar methods have been applied to different problems of the Brownian motion by Smoluchow-

---

[1] *Vide* e.g. B. Riemann-Weber, " The Partial Differential Equations of Mathematical Physics," 4th edit., Book 2, p. 91.

ski,[1] Schrödinger,[2] and Fürth,[3] by the solution of the diffusion equation under corresponding boundary conditions.

(10) p. 16.—The meaning of the probable distribution found

$$f(x, t) = \frac{n}{\sqrt{4\pi D}} \frac{e^{\frac{-x^2}{4Dt}}}{\sqrt{t}}$$

is as follows : one imagines a large number of similar particles accumulated at the time $t = 0$ in the immediate neighbourhood of the plane $x = 0$, and then left to themselves ; now, after a time t such a distribution of the particles is spontaneously established that the relative number of particles between the planes x and $x + dx$ is given by $\phi(x, t)dx$. Here we assume that the particles exert no forces on each other. Such a summation of systems may be called in statistical mechanics a "space-summation." If we now look upon this space-summation as a single system and imagine that a very large number of exactly similar systems are set up, and the same experiment carried out with these, it is asserted that the

[1] The reader will find further information on this subject in the next volume of the Ostwald's " Klassiker," which comprises Smoluchowski's papers on the Brownian movement.

[2] E. Schrödinger, *Phys. Zeit.*, **16**, 189, 1915.

[3] R. Fürth, *Ann. d. Phys.*, **53**, 177, 1917.

mean value of the distribution $\phi(x, t)$ obtaining in all these systems at the time t will correspond exactly to the function $f(x, t)$ in the above formula. The purely imaginative summation used here is called a " virtual summation." For an approximate realization of this one can proceed in such a manner (as the experimenter is generally careful to do) that we make use only of one and the same space-summation, and after carrying out an investigation, this is always brought back to the original condition by artificial means.

But another important meaning can be given to the formula if we consider as our system not, as before, the whole space-summation, but the single particle in this space-summation. Then $f(x, t)$ denotes the probability that the particle has been displaced in the time t to a region between x and $x + dx$. If one observes the movement of a single particle and notices the displacements experienced in successive intervals of time, the relative frequency of these displacements will likewise be given by our formula, in the limiting case of an indefinitely large number of observations. This is what is indicated in statistical mechanics by " time-summation."

Both methods of observation are actually carried out in connection with the Brownian movement, and both lead to the establishment of Einstein's formula. It would appear also as if in this case the mutual substitution of the two kinds of summations were permissible without any further conditions, which is most decidedly not self-evident from first principles. The question as to the exchangeability of virtual and time-summations belongs to some controversial points in the foundations of statistical mechanics. It can be shown that this exchange can always be carried out when a so-called " ergodisch " system is in question ; yet it has not been possible up to now to give a single example of such an " ergodisch " system. The reader will find a comprehensive discussion of related problems in the article on " The Conceptual Foundations of the Statistical Treatment of Mechanics," by P. and T. Ehrenfest.[1] A paper of R. v. Mises [2] introduces a new view-point.

(11) p. 17.—According to the definition of the mean value the mean square displacement is obtained from the expression

$$\bar{x^2} = \frac{1}{n}\int_{\infty}^{+\infty} f(x, t)x^2 dx = \frac{1}{\sqrt{4\pi Dt}}\int_{-\infty}^{+\infty} e^{-\frac{x^2}{4Dt}}x^2 dx$$

$$= \frac{4Dt}{\sqrt{\pi}}\int_0^\infty \sqrt{y}\, e^{-y} dy = \frac{4Dt}{\sqrt{\pi}}\left\{\left[-\sqrt{y}\, e^{-y}\right]_0^\infty\right.$$

$$+ \frac{1}{2}\int_0^\infty e^{-y}\frac{dy}{\sqrt{y}}\right\} = \frac{4Dt}{\sqrt{\pi}}\int_0^\infty e^{-u^2} du = 2Dt.$$

[1] " Encyclopædia of Mathematical Science," vol. iv, 211, part 6.

[2] R. v. Mises, *Phys. Zeit.*, **21**, 225, 1920.

(12) p. 17.—If r is the total displacement of the particle, then

$$r^2 = x^2 + y^2 + z^2,$$

therefore

$$\overline{r^2} = \overline{x^2} + \overline{y^2} + \overline{z^2},$$

and since on account of the homogeneity of the liquid these are all equal—

$$\overline{r^2} = \overline{3x^2}, \qquad \sqrt{\overline{r^2}} = \sqrt{3}\lambda x.$$

(13) p. 18.—The wish expressed here by Einstein was very soon fulfilled, resulting in a complete confirmation of his theory. Amongst the numerous experimental investigations there will only be mentioned here those which have given a direct confirmation of Einstein's formula in its original meaning. The first of these investigations was carried out by Seddig,[1] who took two photographs of an aqueous suspension of cinnabar on the same plate at an interval of 0·1 second, and measured the distance of corresponding images on the plate. He found that on an average the displacements at different temperatures were inversely proportional to the viscosities, as the theory demanded. Henri [2] found similarly with the aid of cinemato-

[1] R. Seddig, *Phys. Zeit.*, **9**, 465, 1908 ; *Zeit. f. Elektrochem.*, **73**, 360, 1912.

[2] V. Henri, *Comptes Rendus*, **146**, 1024, 1908 ; **147**, 62, 1908.

graph records of the mean displacement of particles of caoutchouc that the time law, x^2 proportional to t, was followed.

The establishment of the first complete and absolute proof of the formula lies to the credit of Perrin [1] and his pupils Chaudesaigues, Dabrowski, and Bjerrum, who followed the movements of single particles of gamboge or mastic under a microscope and recorded their positions at equidistant time intervals by means of an indicating apparatus. In this manner they could also use the formula to determine the Loschmidt number N in a new way, and found values between 56 and $88 \cdot 10^{22}$. They could also confirm the distribution law for the probability of different displacements $f(x, t)$ in a quite unexceptionable manner.

Svedberg [2] and Inouye made their measurements in similar manner in metal sols of appreciably smaller particle size, and found a good agreement with the formula with large particles, but systematic departures from it with small particles : this is most probably to be ascribed to a breakdown of Stokes' law in connection with very small particles. They found the time-law well confirmed, and obtained approximately

[1] Perrin-Lottermoser, "The Atom," Leipzig and Dresden, 1914.

[2] Th. Svedberg, "The Existence of the Molecule," Leipzig, 1912.

62·10²² for Loschmidt's number. Similar results were obtained by Nordlund with an automatic-registering photographic arrangement.

Finally, a paper of K. Seelis [1] should be mentioned, which forms a continuation of Seddig's work and amplifies it suitably.

The Brownian motion was first described in gases by F. Ehrenhaft,[2] who has also shown that the order of magnitude corresponds with the Einstein formula. A direct confirmation of Einstein's formula is out of the question here, since the determination of the size of the particles cannot yet be carried out with accuracy ; nevertheless, it appears to be established from the former investigations that the formula can be applied with accuracy in this case too.[3]

(14) p. 23.—The law of Boltzmann mentioned here by Einstein is that known by the name of the e^{-hx}-theorem, which plays a great part in Statistical Mechanics. It leads to the inference that a system which is in statistical equilibrium with another of indefinitely large energy is actually subject to fluctuations, whose relative frequency is given by the law (I). Wherever, therefore,

[1] K. Seelis, *Zeit. f. Phys. Chem.*, **86**, 682, 1914.

[2] F. Ehrenhaft, *Wiener Ber.*, **116**, (IIa), 1139, 1907.

[3] Further details will be found in the report of Th. Svedberg, *Jahrbuch der Rad. u. Elektr.*, **10**, 467, 1913 ; and R. Fürth, *ibid.*, **16**, 319, 1920.

similar equilibria exist, this law should be applied for the calculation of the magnitude of the fluctuations. The reader will find a detailed discussion of all phenomena of this type in Physics in Fürth's paper, " Fluctuation Phenomena in Physics." [1]

(15) p. 25.—M. v. Smoluchowski [2] has given a detailed theory of the Brownian movement under the influence of an elastic force, and, in this particular case, has set forth in a very pleasing manner the points of agreement and differences between the statistical and purely thermodynamic conceptions of natural processes, especially concerning the apparent contradiction between the principal reversible mechanical and the irreversible thermodynamic processes. He pointed out also that this case can be verified experimentally by observation of Brownian torsional vibrations of a small mirror fastened on a thin thread, or of the vibrations of the free end of a thin elastic quartz fibre. The last suggestion was experimentally verified quite recently by P. Zeeman, though finally satisfactory results have not been obtained up to now.

[1] R. Fürth, Vieweg Collection, No. 48, Brunswick, 1920 ; and *Phys. Zeit.*, **20**, 1919.

[2] M. v. Smoluchowski, *Krakauer Ber.*, **418**, 1913 ; Lectures on " The Kinetic Theory of Matter and Electricity," *loc. cit.*

(16) p. 26.—If it be assumed that the energy E of the system is continuously divisible, from the definition of the mean value of a function we obtain for the mean energy per degree of freedom, regarding the energy itself as the parameter α, from the expression (I)—

$$\overline{E} = \int_0^\infty E dw = \frac{\int_0^\infty E e^{-\frac{N}{RT}E} dE}{\int_0^\infty e^{-\frac{N}{RT}E} dE}$$

$$= -\frac{RT}{N \int_0^\infty e^{-\frac{N}{RT}E} dE} \left\{ \left[E e^{-\frac{N}{RT}E} \right]_0^\infty - \int_0^\infty e^{-\frac{N}{RT}E} dE \right\}$$

$$= \frac{R}{N} T.$$

Accordingly the mean energy of a linear oscillator is also equal to this quantity. On the other hand, Planck has shown in the paper quoted that the mean energy of such an oscillator, which is in dynamic equilibrium with the radiation in a hollow body, is given by

$$\overline{E}\nu = \frac{L^3}{8\pi\nu^2}\rho\nu,$$

where L indicates the velocity of light, ν the frequency, and $\rho\nu d\nu$ the energy of that part of the radiation per unit volume whose frequency lies between v and $v + dv$. By equating the two expressions, it follows that

$$\frac{RT}{N} = \frac{L^3}{8\pi v^2}\rho v,$$

and thence the expression given in the text for ρv.

But if it be assumed that the energy E is not distributed to the oscillator in a continuous manner, but only in multiples of an elementary quantum hv, where h is a universal constant, as Planck and Einstein have assumed, there is obtained for E a summation of the form

$$E = \frac{\sum_0^\infty nhv \cdot e^{-\frac{N}{RT}hv \cdot n}}{\sum_0^\infty e^{-\frac{N}{RT}hv \cdot n}}$$

$$= \frac{hv \cdot e^{-\frac{N}{RT}hv}}{\left(1 - e^{-\frac{N}{RT}hv}\right)^2} \left(1 - e^{-\frac{N}{RT}hv}\right) = \frac{hv}{e^{\frac{hv}{kT}} - 1}.$$

As can be seen by expanding the denominator, for small values of v and for high temperatures respectively this formula becomes $E = kT$, which agrees with the former expression. In general, however, by equating the value of E with Planck's value given above, there is obtained

$$\frac{hv}{e^{\frac{hv}{kT}} - 1} = \frac{L^3}{8\pi v^3}\rho v,$$

from which follows

$$\rho\nu = \frac{8\pi\nu^3 h}{L^3\left(e^{\frac{h\nu}{kT}} - 1\right)}$$

which is in agreement with Planck's radiation law.

Hence the " incompleteness of our physical conceptions " perceived by Einstein is related to the necessity for introducing the quantum hypothesis.

(17) p. 27.—This expression can be interpreted again, in the meaning of statistical mechanics, in two ways (*vide* Note 10). If one considers a very great number of particles similar to one another, it gives the relative mass of those particles which will be found on an average at a height $x \ldots x + dx$ above the ground. As the form of the expression shows, the " space-summation " of the particles corresponds with the well-known vertical aerostatic distribution : which is implied by the nature of the case, since only a quantitative, and not a qualitative, distinction exists between the gas built up of molecules and a suspension of microscopic particles. We can now investigate whether a suspension of small, similar particles is actually arranged in accordance with this formula, and, on the other hand, whether it agrees with the absolute figure, i.e. whether by determining the other data Loschmidt's number can be calculated. This was the manner in

which, for the first time, an exact confirmation of Einstein's theory was obtained, a result for which thanks are due to Perrin and his pupils (*vide* Note 13). Their method of procedure was first to prepare, by the device of " fractional centrifuging " worked out by Perrin, a suspension of gamboge or mastic with particles of exactly equal size. This was then enclosed in a microscopic chamber and the distribution in height of the particles determined, after equilibrium had been established, by counting in the microscope the particles in different layers above the bottom of the chamber. In order to facilitate the counting, a small screen was introduced in the ocular of the microscope, so that at all times only a small number of particles were in the field of vision at the same moment ; these were made visible at regular intervals of time by intermittent illumination of the preparation, and so a great number of observations were arranged for. Observations carried out on different sizes of particles and suspension-media showed throughout a very good agreement between theory and experiment, and for Loschmidt's number figures between 55 and $80 \cdot 10^{22}$.

Perrin's investigations have been recently extended by Westgren [1] to still smaller particles of

[1] A. Westgren, *Zeit. f. Phys. Chem.*, **83**, 151, 1913 ; **89**, 63, 1914 ; *Arch. f. Mat. Astr. och Fysik*, **11**, Nos. 8

colloidal metals, which were prepared of equal particle-size by the germ method of Zsigmondy. The method of observation was somewhat similar to that of Perrin ; the results of the extremely accurate investigation were in complete agreement with the exponential distribution in height expected, and give, therefore, pretty well the most accurate direct determination of Loschmidt's number, which was fixed thus as $(60 \cdot 6 \pm 2 \cdot 0) \cdot 10^{22}$.

If the concentration of the particles becomes so great that the forces (operative at a distance) acting between the particles exclude the assumption of the mutual independence of the particles, we must expect divergences from the aerostatic distribution in height. Actually, Costantin [1] found such divergences in gamboge emulsions at very high concentrations, following Perrin's method : which can be accounted for, in the manner indicated, by forces of repulsion between the particles, which are evidently of an electrical nature.

But principally we can look upon our formula in quite a different manner, namely, as a picture of the time-summation for a single particle, if we imagine that a single particle is freely movable

and 14, 1916 ; *Zeit. f. anorg. Chem.*, **93**, 231, 1915 ; **95**, 39, 1916.

[1] R. Costantin, *C.R.*, **158**, 1171, 1914 ; *Ann. de phys. et chim.* (9), **3**, 1011, 1915.

above a floor to which it cannot adhere. Then, as a result of its Brownian motion, it will not simply sink to the bottom, as one might expect, and remain lying there, but will always raise itself spontaneously and dance hither and thither in an irregular manner. The formula gives, then, simply the " relative length of sojourn " of the particle in the different layers x . . . $x + dx$ above the bottom within a long period of observation. This procedure was introduced by R. Fürth,[1] and he was able to show that the formula applies very well to the movement of the particle : here there is further the advantage over the former method that it is necessarily independent of irregularity of particle size and of forces acting at a distance between particles.   It was also possible in this manner, by simultaneous determination of the size of the particle according to Stokes' law, to make a fresh determination of Loschmidt's number :   which gave $N = 64 \cdot 10^{22}$. We see established here in the most pleasing manner Einstein's assertion that the particles can be suspended in the liquid if they be fine enough.

(18) p. 30.—On account of the small magnitude of Δ, we can put the lower limit of the integral in n_3 equal to zero, and develop the function F in the following manner :—

$$F(\alpha_0 \pm \Delta) = F(\alpha_0) \pm \Delta \cdot F'(\alpha_0).$$

[1] R. Fürth, *Ann. d. Phys.*, **53**, 177, 1917.

Thence it follows further that

$$n_2 - n_3 = - 2F'(\alpha_0)\int_0^\infty \xi d\xi \int_\xi^\infty \psi(\varDelta)\alpha\varDelta.$$

If we exchange here the order of the integrations the limits for ξ will be 0 and $\varDelta$, and the limits for $\varDelta$, 0, and ∞, from which it immediately follows that

$$n_2 - n_3 = - F'(\alpha_0)\int_0^\infty \varDelta^2\psi(\varDelta)d\varDelta$$
$$= -\frac{F'(\alpha_0)}{2}\int_{-\infty}^{+\infty} \varDelta^2\psi(\varDelta)d\varDelta$$

on account of the condition that

$$\psi(\varDelta) = \psi(-\varDelta).$$

(19) p. 31.—In the original paper there is given here in error $4 . 10^{23}$: actually the value of Loschmidt's number is $N = 60\cdot6 . 10^{22}$, according to the most accurate measurements that we possess up to the moment.

(20) p. 33. — The formula for the rotary Brownian movement was established in 1909 by Perrin, by suspending spheres of mastic of about 12μ diameter in water and following their rotational movement, as a function of the time, by observation of small, differently coloured inserts in the particles. The formula could thus be closely confirmed, as well as the absolute dimensions : for Loschmidt's number was obtained $N = 65 . 10^{22}$.

(21) p. 33.—If we call e the quantity of electricity that is displaced across any given cross-section of the conductor in time t, and identify α with the quantity of electricity which has flowed across this cross-section since the time $t = 0$, then $\Delta = e$, and α will be the current i. The potential energy corresponding to the displacement Δ is evidently identical with the electrical potential difference, and hence the fictitious force of the E.M.F. E. According to its definition, therefore,

$$B = i/E = 1/w$$

if w indicates the resistance of the closed circuit. We obtain therefore

$$e^2 = \frac{2RT}{Nw} \cdot t.$$

A number of similar questions which are formally related in the closest manner with the Brownian movement have been dealt with by Frau de Haas-Lorentz [1] according to the method of Einstein and Hopf.[2] The list of possible investigations given by her could be considerably increased : however, it has not yet been possible to discover these phenomena experimentally,

[1] G. L. de Haas-Lorentz, " The Brownian Movement and Related Phenomena," *Sammlung Wissenschaft*, 52, Vieweg, 1913.

[2] A. Einstein and Hopf, *Ann. d. Phys.*, **33**, 1105, 1910.

since on account of their minuteness they escape our measuring instruments.[1]

(22) p. 35.—For the lower limits of validity of Einstein's formula and its substitution by a more accurate one valid for any desired small time-interval, refer to Note (8). R. Fürth has also derived an estimate for the lower limits of validity, from the formula quoted above and from other considerations communicated in the same paper, and arrives at the conclusion that this time must be of the order mB, where m indicates the mass and B the mobility of the particle. Actually, one obtains in this manner, i.e. in the order of magnitude, an agreement with Einstein's estimate.

(23) p. 36.—A correction of the following paper appeared a few years later with the title : A. Einstein, " Correction of My Paper, ' A New Determination of Molecular Dimensions ' " (*Ann. d. Phys.*, **34**, 591, 1911), in which some numerical errors in the previous communication were rectified, which had also some influence on the results. In the reprint given here the resulting corrections are already introduced in the text with the aid of the paper mentioned, in order to facilitate the reader's task. The points of correction are indicated in the text by reference to this note.

[1] R. Fürth, " Fluctuation Phenomena in Physics," *Sammlung*, Vieweg, No. 48, Brunswick, 1920.

(24) p. 39.—On account of the incompressibility of the liquid, the " divergence " of the liquid flow must be, on the whole, equal to zero, i.e.

$$\text{div } \mathfrak{u} = \frac{\partial u_0}{\partial \xi} + \frac{\partial v_0}{\partial \eta} + \frac{\partial w_0}{\partial \zeta} = 0$$

or

$$A + B + C = 0.$$

(25) p. 44.—Since

$$\frac{\partial^2 \rho}{\partial \xi^2} = \frac{1}{\rho} - \frac{\xi^2}{\rho^3}, \quad \frac{\partial^2 \rho}{\partial \eta^2} = \frac{1}{\rho} - \frac{\eta^2}{\rho^3}, \quad \frac{\partial^2 \rho}{\partial \zeta^2} = \frac{1}{\rho} - \frac{\zeta^2}{\rho^3}$$

and

$$\frac{\partial^2 \frac{1}{\rho}}{\partial \xi^2} = \frac{3\xi^2}{\rho^5} - \frac{1}{\rho^3}, \quad \frac{\partial^2 \frac{1}{\rho}}{\partial \eta^2} = \frac{3\eta^2}{\rho^5} - \frac{1}{\rho^3}, \quad \frac{\partial^2 \frac{1}{\rho}}{\partial \zeta^2} = \frac{3\zeta^2}{\rho^5} - \frac{\rho^3}{\rho^3}$$

we obtain in (5a)

$$\frac{\partial D}{\partial \xi} = \frac{5}{6} P^3 \left\{ \frac{\partial \frac{1}{\rho}}{\partial \xi} (A + B + C) + \frac{3\xi}{\rho^5}(A\xi^2 + B\eta^2 + C\zeta^2) - \frac{2\xi}{\rho^3}A \right\}$$

$$+ \frac{1}{6} P^5 \left\{ - \frac{\partial \left(\frac{1}{\rho^3} \right)}{\partial \xi}(A + B + C) \right.$$

$$\left. - \frac{15\xi}{\rho^7}(A\xi^2 + B\eta^2 + C\zeta^2) - \frac{6\xi}{\rho^5}A \right\}$$

Since $A + B + C = 0$, two of these terms vanish, and the remainder, put in u (5), gives

$$u = A\xi - \frac{5}{3}P^3 A \frac{\zeta}{\rho^3} - \frac{5}{2}P^3 \frac{\xi}{\rho^5}(A\xi^2 + B\eta^2 + C\zeta^2)$$

$$+ \frac{5}{2}\frac{P^5}{\rho^7}\xi(A\xi^2 + B\eta^2 + C\zeta^2) + \frac{5}{2}P^3 A \frac{\xi}{\rho^3} - \frac{P^5}{\rho^5}A\xi$$

from which equation (6) follows immediately.

(26) p. 48.—A remark is omitted here which refers to the amount of energy consumed, since this is no longer involved after correction of an error of calculation (*vide* Note (23)).

(27) p. 53 —Following from Gauss' law

$$\int \text{div } \mathfrak{u} \, do = \int u_n ds \, ;$$

since u lies in the direction of the x-axis, and in sum-total is equal to u, then

$$\text{div } \mathfrak{u} = \frac{\partial u}{\partial x}$$

and

$$\mathfrak{u}_n = u \cos (x, \, n) = u \frac{x}{r}.$$

(28) p. 62.—The values given here for the radius of the sugar molecule and Loschmidt's number agree remarkably well with determinations of these quantities made in other ways.

The most accurate value for Loschmidt's number at the moment is $6 \cdot 06 \cdot 10^{23}$, determined from the " Faraday " of electrolysis and Millikan's value for the elementary quantum. The values, from the Brownian movement, given in the preceding notes, agree, therefore, remarkably well ; as well as that derived from the radiation of heat according to Planck's equation, $64 \cdot 10^{22}$. Further, from the Einstein[1]-Smoluchowski [2] theory of

---

[1] A. Einstein, *Ann. d. Phys.*, **33**, 1294, 1910.

[2] M. v. Smoluchowski, *Ann. d. Phys.*, **25**, 205, 1908.

density-fluctuations in gases and liquid mixtures, a value can be obtained for Loschmidt's number from measurements of "critical opalescence," in gases in the neighbourhood of the critical temperature, in liquids in the neighbourhood of the critical miscibility point. The former observations were carried out by Kamerlingh Onnes and Keesom,[1] and gave approximately $75 . 10^{22}$, the latter by R. Fürth[2] gave $77 . 10^{22}$, and by F. Zernike[3] with more accurate equipment, 62 to $65 . 10^{22}$. According to this theory, Loschmidt's number can also be determined from the extinction-coefficient of air for sunlight, by which method Dember[4] obtained $64 . 10^{22}$. We see, therefore, that a very large number of completely independent methods exist which all lead to approximately agreeing values for this important constant.

With regard to the size of the sugar molecule, there is available for comparison the diameter of the first electron ring of hydrogen, derived from Bohr's theory of the hydrogen spectrum, about $0·5 . 10^{-8}$ cm., whilst the sugar molecule would be about 100 times as large as this, the smallest of the atoms. According to the kinetic theory of

[1] W. A. Keesom, *Ann. d. Phys.*, **35**, 597, 1911.
[2] R. Fürth, *Wiener Ber.*, **124** (2a), 577, 1915.
[3] F. Zernike, *Dissertation*, Amsterdam, 1915.
[4] H. Dember, *Ann. d. Phys.*, **49**, 590, 1916.

gases, the diameters of gas molecules are of the order of magnitude 10^{-7} cm.

(29) p. 63.—Th. Svedberg, "On the Spontaneous Movements of Particles in Colloidal Solutions." First paper, *Zeit. f. Elektrochem.*, 12, 853-860, 1906. Second paper, *Zeit. f. Elektrochem.*, 12, 909-910, 1906.

(30) p. 65.—From

$$m \frac{dv}{dt} = - 6\pi kPv$$

it follows that

$$- \frac{m}{6\pi kP} \frac{dv}{v} = dt$$

and by integration

$$t = - \frac{m}{6\pi kP} \log v + \text{const.}$$

or, since when $t = 0$, v will equal v_0

$$t = \frac{m}{6\pi kP} \log \frac{v_0}{v},$$

from which it follows, for $v = v_0/10$

$$\theta = \frac{m \log 10}{6\pi kP} = \frac{m}{0 \cdot 434 \cdot 6\pi kP}.$$

(31) p. 66.—See, e.g., the treatment in Section 1 of this volume on p. 17.

(32) p. 70.—Compare, e.g., the section, "Osmotic Theory," in W. Nernst, "Theoretical Chemistry" (Stuttgart), under Electrochemistry.

(33) p. 75.—Compare Section 1 of this volume, paragraph 3, p. 12.

(34) p. 76.—With reference to the Note on p. 75, the uncertainty in the value of N indicated here corresponds with the contemporary state of progress in the enquiry. To-day this uncertainty can be put at scarcely more than 2 per cent. *Vide* the agreement in the different methods for the determination of N, Note (28), p. 116.

SUBJECT INDEX

8 *

AUTHOR INDEX